ISBN 0982737106
ISBN 978-0-9827371-0-1

First Published during February 2009, as a Digital Web Book.

$19.95
ISBN 978-0-9827371-0-1
51995>
9 780982 737101

History has proven;
The world is filled with terrorists, drug cartels, drug dealers, drug users, predating pedophiles, stalkers, rapists, good cops, corrupted cops...

The list may go on to even more monstrous cases...

INSIGHT IGNITES THE EYEWITNESS

Book One
SITUATION RESOLVED

By
Mr. Cloak

As we took a few steps closer, my eyes met the eyes of the older of the two men. Instantly, my neck straightened, my eyes opened wider, my inhaling of breath paused. Without giving it a second thought, I knew him! "When I looked at his face, into his eyes, it was instant recognition, and he knew it! I had seen him before." It was "Osama bin Laden". Instantly, his neck straightened, his eyes opened wider, his inhaling of breath paused. Without giving it a second thought, he knew I recognized him!

To:

My family and friends, who accepted, some with apparent resentment, and more than audible complaint; Myself-isolation during the many years that have passed, sense the end of the Gulf War....

Open your eyes widely!

To:

Joe Alercia, Dean and Founder of "Lion Investigation Academy." I cannot imagine, how badly it could have turned out without the knowledge I gained from your school...

Well then, it is time to put them three steps behind…
Joe Alercia II – 2003…

Joe, you are the man...

To:

Senator McCain and all members of his wonderful staff; If American's only knew 50% of the total truth, of what goes on covertly behind closed doors, and around the world to protect us! You Sir; would have been elected! As the most popular President of all time...

I'd rather lose an election; then lose a war!
John McCain - 2008...

Sir, you inspire me...

"Capturing the Essence"
This Book Is Dedicated; To My Grandmother,
Whose First Name Was, "America"?
She grew up knowing survivors of the "Civil War"!
GOD BLESSES AMERICA, NOW AND FOREVER!

CONTENTS:

FORWARD

INSIGHT IGNITES THE EYEWITNESS

Book One

SITUATION RESOLVED

An Eyewitness Account of "Osama bin Laden",

By An Unwanted, Unappreciated Combat Wounded Veteran, Of The 82nd Airborne, United States Army!

"In this recession" why are the Philippines getting your tax dollars? Smile, I'm going to spill the beans on this one….

They're camouflaging it, but I'm not! They're trying to hide it, but I'm not! They're not even talking about it, but I am!

This is not a conspiracy! What this is; is an eyewitness account, a personnel journal! What you are about to read, I have seen with my own eyes and experienced with my own being!

"When I looked at his face, into his eyes, it was instant recognition, and he knew it! I had seen him before." It was "Osama bin Laden!"

"This Is Extraordinary Info!" Get ready to become totally super motivated. If you do not want to know how, and where, I walked up on "Osama bin Laden", then you are absolutely totally and completely crazy! As long as there is "Islamic Extremist", your life is in danger….

I walked up on "Osama bin Laden," it is time I let all of you know what took place. Before, during, and after, even until this very day! I will show you copies of the "SALUTE REPORT," I sent by email to President Bush, Vice President Cheney, the CIA, the FBI, Senator McCain and more, concerning my encounter with "Osama bin Laden."

As you might imagine, due to the need for personnel security, I will not use my real name, or the names of my loved ones, in this "Eye Witness Account".

Further, I reserve the right to sue in all courts around the world, anyone that discloses my personnel information in any way, shape or form... Even the President of the United States on the grounds of "Gross Deliberate Endangerment of Life," by use of a "Class Action Law Suit!"

You will come to know me, simply as "Mr. Cloak", my wife as "Hun," my son as "Bubby" along with my adopted daughter as "My Little Angel!"

In this book, I am going to inform you as to my encounter with "Osama bin Laden," the days before, and the days to follow. I will make it simple, informative, direct and to the point. However, at the beginning of this book, you will get to know me, maybe even identify with me; chances are you are very much like me. Then at the end of this book, you will know no more than I do and the reason why!

The Cover Up..................

Take out a pen and pad of paper; start taking note's right away! Start your investigation today!

In "**SITUATION RESOLVED**" you will learn how I found "Osama bin Laden," who I contacted and why! From this day forward, there will not be a day that goes by, when you do not look around you with a wide-open eye. You will know what I know, and how to discover even more!

You will understand that the "Cold War" never ended, that the KGB only moved into a different building, and changed their name. Look around you, North Korea now has ballistic missiles that can reach the continental US which carry nuclear bombs. Iran is building them also. Will they sell them to Venezuela, and Cuba? Alternatively, will they simply give it to them!

Just a few short chapters of this "Eyewitness Account" and you will be voting in future election's with total conviction, for years to come. You will even work toward getting every member of your family involved, and voting down government run health care!

Nevertheless, you will also realize how to become a source of information for local crime fighters, even how to get your information to the right people, and used by the most well known "Anti Terrorist Organization" worldwide.

All you need to be able to do, is see the obvious, while acting like, you do not know what is going on.

And you will, for I will teach you how! You are now, a volunteer undercover operative.

You will see the world around you make changes, which you brought about. I am going to show you, how I do just that. Come join me, as we "RATS PATROL"!

You will soon want to have me publishing book after book that "All Freedom Loving People's" simply must have. Although, I reserve the right to say "NO!", and you will soon know why!

There is still more! When a government official comes to your town, you will be there, at the "Town Hall Meeting," to tell them exactly how you expect them to do their job.

Okay, I could go on! There is literally no end in sight with what I need to tell you! So open this book and read, about how this "Situation Evolved," and how it was "Resolved."

Whatever outrageous, positive or negative thinking you believe in, or if you are even remotely interested in the war on terror. You would be absolutely crazy not to read this book.

You can change the world!

This eyewitness account will actually exceed your expectations...

Life has thrust me into a world of undercover operatives, Without pay!

SYNOPSIS:

INSIGHT IGNITES THE EYEWITNESS

Genre, War on Terror, ***Historical Non-fiction!***

Book One
SITUATION RESOLVED

An Eyewitness to "Osama bin Laden," has come forward: Together with his wife, a retired D.O.D. operative sits down at a table outside of a coffee shop. At the next table, sitting among the group is a clean shaved man; he knows to be "Osama bin Laden!"

If he reports it, trembling in fear, sure they will be killed. His wife claims she will divorce him! Danger in her life is a hell to hide from at all cost!

Spiraling through an insane situation, he knows he must contact a higher source! As his many years of experience ignite, he realizes he is being stalked.

At night, dressed in a black military uniform, sterile of any insignia, a man walks by him on a crowded street and blatantly says;

Follow You Die!

With the eyewitness account of Osama bin Laden, firmly locked inside his recent memory, this can only mean one thing: Osama bin Laden, knows he has been discovered, and, has sent his assassins!

As the world wonders, where is Osama bin Laden? This old timer has informed the CIA along with other US Intelligence Agencies of the US government: The activity and location of the world's most wanted terrorist, the leader and founder of al-Qaida!

The days that follow have many dangerous, and or historical events; with the top secret knowledge of his encounter with Osama bin Laden, and, his reporting it. He knows, that the historical events symbolically point to covert searches, or, the capture of Osama bin Laden by CIA operatives! However, covert searches for him by assassins sent by Osama bin Laden are or may be, around each turn!

Now he is followed by operatives on all sides of the war on terror, life has thrust him back into a world of undercover operatives!

Continuing to report all situations to a trusted Senator, it is not until he accounts being interrogated by a Foreign Military Officer, who has infiltrated the US Veterans Administration working as a doctor, does this climactic moment, resolve his dispute with the U.S. government over a decrease in his medical retirement benefits! However, this ignites a whole new perspective:

We have all been betrayed!

Now his young son is being followed. Again a stranger strolls by on the street, to claim bluntly:

You Will Die!

He has heard no fat lady singing, the saga continues, building toward a, **Book Two!**

To Date: I have successfully eluded, all assassins sent by Osama bin Laden!

Both sides follow me... I am the eyewitness....

Special Note: From time to time, in this book, you will read copies of authentic emails sent to important people. You will find that the author's email address and name is altered to protect his identity. When needed, the name of loved ones will also be altered!

The names of other, real life individuals have also been changed to protect the innocent!

If at any point in this book, you find information left un-altered, that can lead to the identity of the eyewitness. Please contact the publisher…

INTRODUCTION:

The benefits of a retired D.O.D. operative, are not very lucrative, at least not when they are awarded after a long and painful court battle.

You are not respected!

Recounting and re-living torture or gunshot wounds, including the horrific death of your comrade in arms, are stressful events!

What do they want me to do, spill the beans? How could the government of the U.S.A have gone so far down the tubes?

My years in the U.S. Military, have left me with many experiences. As you read forward, notice that I am writing this book, as if I were talking to my best friend, that's you!

Get ready to become super motivated. If you do not want to know how and where, I walked up on "Osama bin Laden", then you're just plain nuts! As long as there is "Islamic Extremist", your life is in danger. You know it's true!

On the morning of 911, I was awakened by the father of the woman; I was married to at that time. I had been sleeping on the couch in the living room of my house all night, in Florida, as he knocked on the front door. Came in and said, "Turn on the TV, we are at war!"

I had spent the night on the couch, because the night before as my wife at that time, walked through the front door, after arriving home late from work! I attempted to give her a welcome hug and kiss! While doing so, I smelled on her breath what I believed to be, bodily fluids of a man.

Two and Two Equals Four!
Connect the dots…
I had been betrayed!
Life can seem to be cruel, can't it!

This sent me to search out a study course in private investigation. Doing so, led me to Joe Alercia and his "Lion Investigation Academy" of Bethlehem Pennsylvania, USA. Toll Free: 1 (877) PI School. Where I earned an "Associate Degree in Private Investigation and Security"! This reawakened in me, a deep desire to seek out and live in the truth.

After months, and an existential investigation of which crack cocaine was found to be the main cause! It had become painfully obvious that many of my fellow man, black, white and brown - once they received word of it, were willingly standing in line for their turn with this man's wife. This list includes neighbors, fellow past class mates, strangers and even members of the local law enforcement offices, lawyers and judges not excluded.

It was not the color of their skin that was to blame. It was simply the "Content of Their Character." The FBI confirmed that she not only was hopelessly addicted to crack cocaine, but that she was also a "Border Line Personality!"

I would eventually turn my back on this, and walk away!

But for the time being, this sent me spiraling into a living hell, for I had been here before after returning from the Gulf War during 1991.

This time, my living hell lasted until mid 2003, in the country of Panama. This is where I was wounded in combat years before as a member of the 82nd Airborne US Army!

It seemed as though I was under some kind of evil spell, so I had gone to the Veterans Administration to seek counseling.

Sometimes, it is best to talk to a stranger about personal problems in your life. Preferably, someone that you trust not to seek you out! Alternatively, even, someone that you will not run into on a day to day basis!

During counseling with a doctorate holding professional, I brought up the fact that I had been in the same type of situation after returning from the Gulf War, except this time crack cocaine was involved.

And, the more I had talked about it openly to friends and family, the worse the situation became! At that time, the counselor looked at me directly in the eyes and said;

"I want to help you!"

"Here's what you have to do!"

"When you see these things happening around you,
Act like; you do not know what is going on!"

"Do not speak about it to anyone!"

"Do not talk to family or friends about it!"

"Only tell the police!"

(Further in this book you will find that this technique, once again, would help out immensely!)

This technique worked excellent, and eventually helped me to let go of my loved one, and move on. The only problem I ran into was the fact that, inside the local Sheriff's Department an internal war was under way!

To the extent that certain deputies were being played by the women in the drug culture, and withholding information from the detectives department. The whole time helping individuals in the local drug culture, as they attempted to find a way to bring me down after they had become knowledgeable about the fact that I was a Volunteer Undercover Informant!

A Sheriff's Deputy and old class mate even went to the extent of, during off-duty hours and in civilian clothing, went around ahead of me. Trying to manipulate the situation to where I could be framed!

Fortunately for me, my eyes were wide open, and I had found within the sheriff's detectives department, an honest man who had, like me, served in the United States Army as an infantryman! Although again, it seemed as though the devil had found a way to get at me and make my life a living hell!

To break the spell that had befallen me, I freely choose to go back to the country of Panama and face a demon from my past.

This long awaited confrontation benefited me significantly: I have moved on!

Please make a note of these subsequent circumstances: If you find yourself answering the phone to hear a person of the same sex as you; asking for a person of the same sex that you are. However, with a strange name, that you are not familiar with, begin your investigation. As it is highly possible, through my first hand knowledge, that an ill reputed sex club has been successfully utilizing a coded message system against you!

By which they contact members to send a needed message. In this case, to your spouse or partner, although this could be used for anyone in your family!

Take the following incident that happened to me, put it to memory, then utilize it to your advantage, friend!

Recognize that I, was acting as if I did not know what was going on; whilst I had begun to notice that my wife at that time hurriedly set about gaining possession of our phone bill. Once she had control of the phone bill, the necessary payment was made, then it was put into the garbage.

While unaccompanied; I went to the garbage looking for the old phone bill. When I had retrieved it, I noticed twenty nine separate phone numbers that had called my wife's cell phone. She also had called them on many occasions during the week, while she was at work!

When tracing the phone numbers, I found that they were at locations throughout our county and our neighboring county. These were locations that, she, and I had never visited together or spoken about!

At one time while she was in the backyard, her cell phone rang. As I answered it, a man on the other end, said "Ronnie it's time to call your brother, MAN!"

My name is not Ronnie! Also, there was something about the way he delayed when he heard my voice. He then made the announcement, hanging up the phone.

The way he also broke off before he improvised the word "MAN!"

Later, I was able to deduce that each member of the sex club, had a code name; a common name, which was given as if to someone of the opposite sex than they actually were!

You can see how this can work easily either way, whether you are male or female!

Example: let's say you're a man, in the sex club you will have a female's name as your code name. This way, when someone in the club tries to call you! However, your spouse answers the phone, or the answering machine picks up the call; the club member uses your code name, which is the name of a woman.

They go on to make an announcement of some kind referring to the opposite sex then you are (Which is a coded message to you!). This throws off anyone from completely understanding that the phone call was meant for you, a man!

The same goes if you are a female! In the sex club, you will have a man's name, as your code name. This way, when an individual in the club tries to call you but your partner answers the phone. Or the answering machine picks up the call! The club representative can use your code name, which is the name of a man. Providing you with a very convincing argument, that it was simply a call to a wrong telephone number!

They go on to make an announcement of some kind referring to the opposite gender then you are. This throws off anyone from completely understanding that the phone call was meant for you, a woman!

Now read that once more! And recognize how easily you would disregard it as a telephone call to a wrong number. In this situation, you did think that somebody had called a wrong number. Moreover, when they realized it had just hung up the phone!

Otherwise, you hear the same message, or type of message left on your answering machine or as a recorded message on your partner's cell phone. How comfortably it would be for you to just disregard it as a telephone call to a wrong number!

In my situation, it even deceived the local sheriff!

Now, if you are single and have the chance to meet an important person! In spite of this, it turns out that they have been divorced. Do not automatically assume that they were the offender. They might have been victimized, extremely!

It does not matter if they are male or female! They may have been the victim and deserve your consideration! So do not jump to judgment of a divorcee, be prepared to give respect where respect is deserved!

It is a disadvantage, in our day and time, even as Americans or citizens of the free world! Simply, because, in many cases marriages are only a few words on a piece of paper! It does not truly mean the person was married, not in the true sense anyway!

They may have been wedded on their side of the agreement only! Meaning, that they were married to their spouse, but their spouse was only playing them for a fool!

Search your spirit and you will know that just because a piece of paper says somebody was divorced. It does not mean that they were ever wedded, not a true marriage, in the true sense of the word!

Mean while, deep inside the ancient corridors of my mind, the memories of a war in a far away land, push their way into this present time. You see, persistently after the "Gulf War" of 1990 and 1991. I have told a version of an encounter with a man I have always thought looked like "Osama bin Laden" on or about Aug 10, 1990 inside Saudi Arabia!

Further in this book (Dec 2007), you will come to appreciate why I have added this at the beginning.

Being a card carrying member of the 82nd Airborne, I was told to jump from military aircraft often. It was my job! On Aug 7th 1990, I and many other paratroopers were loading a "C-141 Star Lifter" for a night jump.

I was nervous! As it had not been many days sense my medical profile had been lifted. After spending months on crutches from being wounded in combat, in Panama City, Panama. During "OPERATION JUST CAUSE"!

Then shortly after, my medical profile had been lifted, having to go back to a medical profile and crutches because of a bout with blood poisoning!

This second term on a medical profile and crutches had only ended a few days before!

However, that's another novel to write, so let's move on!

The crutches were needed to help me walk, because of a gunshot wound by an enemy soldier in a sniper position. Which entered the upper inside of my left leg, and exited my outer left thigh and lower left hip?

A nasty flesh wound that hurt like hell; I literally blacked out!

After surgeons had removed the dead, and or, injured flesh, from the exit wound! A round hole large enough for a grown man to put his fist into remained! It was left to drain for days, before it had to be elongated before it could be sewed together! By cutting 2 large V-shape cuts, at the left most, and right most, portions of my gun shot exit wound!

This had to be done, because the perfectly round hole, cut by the surgeons could not be closed. The 2 large V-shape cuts which elongated the exit wound, allowed the surgeons to sew shut my surgically enlarged, gaping war wound. However, this left me to hobble around on crutches for months to come.

This brings me back to the night, of Aug 7, 1990 and a "C-141 Star Lifter"! A "C 141 Star Lifter" is one of those large gray colored, jet powered aircraft that you see paratroopers jumping out of!

I had worked all day, starting out the morning with the usual hour of physical exercise from 6:00am until 7:00am. I then spent the day doing various tasks, assigned to me as a non-commissioned officer or "NCO", E-5P type, United States Army issue, 1 each of the Headquarters of Headquarters Company, 325th AIR…

At 5:00pm those of us assigned to jump that night reported for pre-jump accountability. We went through pre-jump procedures and loaded the trucks. Moved on to "Green Ramp at Pope Air Force Base", and strapped on our parachutes.

As a qualified jumpmaster, I and other jump masters preformed "JMPI" (Jump Masters Primary Inspection) of each jumper. We then loaded the "C-141 Star Lifter," packing ourselves in like, "Sardines in a Can!"

As we lifted off, and flew our fight pattern. It soon became increasingly hotter. With the flight captain saying over the speaker, that the air conditioning was not working, and because the inside temperature of the "Star Lifter" was reaching 115 degrees, we must all drink water.

Those of you that have served in the military airborne community know what I mean, when I say; "That is not an easy task!" Because, being strapped in and packed tightly, to fill the aircraft with as many airborne soldiers as possible. It leaves no room for hand and arms movements, down and around to your back side, where your water canteen is located!

Imagine if you will the deafening hum of 4 jet engines!

You and 100 or more other human beings are jam-packed like sardines in a can, with the temperature cooking at 115 degrees. On your left side is a professional paratrooper overcome by the high temperature, vomiting into a brown paper bag!

To your right is a trooper silently praying to his God!

Directly, across from you is the face of a person you have never seen before. He is turning grayish in color and tears are filling his eyes.

To his left and across to your right, is a young man passed out, head leaning toward the front as if sound asleep. To his right, across from you on the left, is an unfortunate soul, with one of the load crew members of the aircraft standing on top of his chest! Since there is nowhere to walk except on top of us, while he carries a small paper cup of water. To the professional paratrooper, vomiting into a brown paper bag, next to you!

Now imagine the fresh hot odor of the vomit, as it enters your nose, cascading along your senses. Overwhelming your DNA, with the need to vomit violently all over yourself! You taste it! It is just about to blow; now you are there!

After an hour and a half of this, the jumpmaster calls out, "3 minutes" as loud as he can. You realize that you might live. You reach to the buckle of the strap that holds you down. You yearn to stand up, but know, that each parachutist must wait until the person next to them stands up first. Starting with the trooper nearest the door on your side of the aircraft! Then moving closer and closer to you!

You wait for your turn anxiously. Hoping, the opening of the door will send a blast of fresh cool air to your lungs. You strain forward, for a glance toward the door. You see figures of men; they lean toward each other and shake their heads from side to side, saying "No!"

Over the speaker, the flight crew informs everyone that the door on your side of the aircraft will not open. They will circle around for another approach to the drop zone, while the load crew works to release the door!

After the 3rd pass, the door is still not open. The loud speaker calls out in the fight captains voice and says, that the jump has been canceled, you are being returned to "Pope Air Force Base", for parachute turn in.

Arrival in one hour!

After parachute turn in, I drove home and arrived at 23:30 hrs. I had just taken off my boots, when the phone rang! The voice at the other end said, "Sgt., Full Muster Alert, report to the arms' room, draw you gear. Prepare for immediate deployment, time of call out 22:00 hrs."

As I hung up the phone; I already knew, that 2 hours, was all the time allowed. After a notification to report for duty, before you are subject to the "Uniform Code of Military Justice" and prosecuted!

Looking at the wall clock; it is 23:35 hours… I was being notified an hour and a half after my unit's notification. I had just 25 minutes to put my boots back on, kiss the wife, hug my little baby boy (Bubby) and make the 30 minute drive to my unit's arms' room. Draw my weapon, and dawn my equipment!

With tires squealing, emergency flashers blinking, I did just that, and arrived in "Saudi Arabia" at 13:00 hours on Aug 9, 1990, with 9 other paratroopers from my unit. The rest would be landing later, because they were held over in Spain while refueling.

Did your eyes just open widely?

They should have; two and two equal four!

Those in charge of the scheduled jump knew we were going to be called out way back before the time to open the door of the C-141 Star Lifter! The air conditioner not working was a lie. In reality, it was a test to see how many of us would pass out from the heat!

Does, CIA come to mind?

Enough of us had obviously passed the test!

That explains the many times an unmarked, black helicopter landed at the Brigade Headquarters' building of the 325th A.I.R.!

Stepping out of the C-141 Star Lifter into the hot afternoon air of Saudi Arabia was like stepping into a hot oven, preheated and set to cook at 130 degrees!

No clouds, no shade, on a concrete runway many miles from even a lonely camel!

We were quickly loaded onto a small bus, and then driven through the desert to a makeshift Arabian tent compound. We were told to stay there, while we waited for the rest of our unit to arrive!

Later that day, and all night long, the buses were stopping. American soldiers were getting off, walking up to a tent, with everyone taking off their BDU shirts. The 130 degrees in the sun did not cool quickly after the sun went down! In fact, the temperature never lowered more than a few degrees all night.

Around 3:00pm the afternoon of Aug 10th, overwhelmed by a sense of professionalism! I walked over to a concrete wall at the edge of the makeshift Arabian tent city. Finding an opening, I walked through!

There before me was a three story building! Much like a motel, the type you see on the side of a highway or a school. With walk ways in front, along the whole front side of the building. There in the center divider, was a walk way to the back, with no people in sight.

Walking over to the right front side of the building, I checked a door, then another! They were unlocked, with completely empty rooms, not even a chair! Totally void of any sign, that someone had ever occupied the rooms throughout the entire building.

As I continued to walk toward the end of the building; I came to the stairs, then walked up them to the 2nd floor! Opening each door and looking in; I felt as though this must be where we were going to be housed or where our command center would be located.

As I came toward the center of the building, I opened a door, to find surprisingly, that it had a wall just inside. I walked further in and around the wall. It was a shower room!

Large enough for a foot ball team to shower after a long game. As I tested each shower, finding them to be in working order; I intended to go back to my unit and inform my chain of command. Of the shower I was going to take, then invite all to do the same. Suddenly, I heard the sound of someone talking loudly outside!

As I walked outside, the voice seemed to be further away! Walking back down the stairs, then down the walkway, I came to the hallway that led to the backside of the building!

While walking through the hallway, the voice kept getting louder and louder with each step I took! It felt wrong, the language was not recognizable!

I was suddenly shocked into reality, as I came to the opening at the backside of the building! Own my left, about 50 feet away! Was a group of 30 or so men, listening to a man that was on a small platform? His head was 3 or so feet above there's.

He was exclaiming loudly, something in their language, which had them all riveted! As passionate in his speech as an old time "Baptist Preacher", intent on exercising demons out of his congregation! By enticing the fear of God, into all who heard him!

At that moment, he saw me, turning his head in my direction! He spoke in perfect English, but with an accent saying, "We are being told, that they have sent their son's to our land to protect us! We have been told, to feed and shelter them. To treat them with respect, for they are not barbarians!"

As he spoke, many of the men in the crowd turned their heads, following his gazing direction. They all looked straight at me, then back at him; as if they expected him to give them orders to capture or kill me!

I felt the hate in their eye's, turned to leave, heading directly back to the tents, in my unit area. The whole time, thinking to myself. That, the words he spoke in English, did not match the hate in the eyes of the men listening to him. Nor their actions; He must be saying something totally different to them, in their language!

Once back at my unit; I located my chain of command. Informed them of the showers, and the crowd of men behind the building! Even the speech that they were being spoken to about, by a man that looked like "Bin Laden"!

I was ordered not to return to the area, and not to speak of it to anyone!

The man speaking to that crowd was "Osama bin Laden"!

Two and two, do equal four, connect the dots; this man has been protected by the King of Saudi Arabia, for a long time!

A day later, all of us that had been there since the first day, were now out of our three days worth of MRE's. That was packed in our backpacks. For the next 2 months, all we were given to eat from the Saudi government was "One Hot Dog Bun, three times a day."

By the time October came around; we were receiving one boiled egg in the morning along with our morning ration of "One Hot Dog Bun, per Individual." Just as suddenly as it began, it ended. Only then, did we begin receiving shipments of MRE's.

Two and two do equal four!

Connecting the dots; Although we were there, to protect them with our very own lives! They hated us, to the extent of wanting to see us dead!

Chapter 1

Arrival

My son (Bubby) and I arrived in Manila Philippines during Feb, 2007 at 3:00 AM in the morning! Just off of the plane we had to go through customs. I had given "Bubby" a very nice "Meade Telescope" for Christmas and because one of our family members had sold every personal item we had left behind when we went to Panama. He would not leave it behind in fear that the same fate would befall his "Meade Telescope."

At the custom's desk, a Filipino man wanted to confiscate Bubby's "Meade Telescope," saying that it could be used to spy on the "Philippine Military." I was faced with the stupidity of this man, until a Philippine woman working at the same custom's desk spoke up by saying loudly, "He is a student! All that is used for is watching the moon and stars. Haven't you ever been to school?"

I then had to pay the import cost of the telescope. Before we could meet our ride and preceded to a valley, in the province of Bataan, Philippines. That I had never heard of before my correspondence began with "My Hun".

Within 3 days, feeling so over-whelmed with culture shock, and easily angered, on an hour-by-hour bases. I found myself wishing I could snap my fingers and change the world.

As you know, some people in our world have nothing, have always had nothing! And, will always have nothing! Life has knocked them down, generation after generation. Until they have no hope, or even the desire to get up and leave there none prosperous life behind. To look in a new place, for something just a little better! For them, a bowl of rice is their treasure! This is the way I found the people of this little valley community!

"My Hun" had been living here with her aunt, for the last 8 years of her adult life. She was ready to leave it behind. I was ready and willing to take her away from it.

You see, I am a combat wounded veteran of the US Army. I am living off of "VA and Social Security Disability Compensation." In the USA, this is not enough to do much more than sit home and pay bills.

Here in the Philippines, it is a treasure to kill for. As the daily income of a well trained worker might reach $4.50 per day. Overtime included! Now imagine the lower income of an un-trained worker, and you have a more accurate perspective on this situation.

Take a professional card carrying security guard I know in the city of Olongapo Philippines. He claims that he makes 221 pesos per day. Working a 12 hour night shift, 7 nights a week!

At the time of this writing, that converts to $4.53 per workday. And $135.90 a month, during a 30 day month! This is all that supports him and his family. Even with the price of everything, the same as in the U.S.A... With, the price of gasoline over $4.00 here, long before it went up in the states!

So you can understand that I am looked at as a rich man and rightfully so! For, I am an American, Blessed by God's loving grace, to have when others have not!

The only time in my life, where I have truly felt hunger, was during 1984 while I was training in the US Army SFQC (Special Forces Qualification Course) survival course. Where I was left in the "Uwharrie Forest" for 7 and 1/2 days with a small, white rabbit, "maybe 8 weeks old" for food! A knife, a book of matches, 30 feet of parachute cord, 6 sheets of paper with a broken pencil! Being graded on the many tasks, spoken about during the classroom discussion's days before, on the day they, the Green Beret instructor's, where to pick me up.

I passed the course, but only because I killed that little rabbit and ate it, along with completing the many tasks. That proved I could live in the wilderness like a pre-historic cave man, remaining un-seen and un-known to enemy forces! Until being reunited with the American military.

My friends and I had nicknames like Bum, Conboy, Becker, Brown, Charlie Eight Two and Cold Steel!

You see, Rambo's wilderness survival skills are fictional, because he is a fictional character. My wilderness survival skills are real, proven and documented with the United States Military Special Forces Command. So do not seek me out, I wish to remain anonymous.

Getting married, and being a combat wounded American; I needed to inform the United States Veterans Administration of my recent marriage. To do this, "My Hun" and I had to rent a car, and then travel to the nearest United States Embassy. This was located in the Philippines capital city of Manila!

Special Note to Reader: During this time period, the Philippines were going through an election period and were experiencing many political murders, which continued to interrupt the car radio, to include the TV news channels! "Bubby and I" often pointed out to each other, how people in 3rd world countries actually listen to politicians when they said "I did this or that for you. You have to vote for me!"

We were watching a Mayor buy concrete, and surface a neighborhood street, "2 weeks" before the election. While walking around saying "Look what I did for you! You have to vote for me!"

Nothing had been done for the neighborhood sense the last election.

She was re-elected?

Did your eye's just open wide?

They should have!

Two plus Two, equals Four!

Connect the dots…

The drive was hectic and extremely aggravating, due to the never ending amount of pedestrians darting across the highway, mile after miles! Including the hundreds, even thousands of motorcycles with a sidecar, affectionately called a tricycle by the Filipino people. Which were darting in and out of traffic, seemingly from every direction?

Although my years are filled with many experiences, I was not prepared to handle the four-hour drive at a snail's pace through darting pedestrians and motorized tricycle's laden with local people going about their day to day life.

Not to mention the enormous strain on the scar tissue of my war wound, from continuously pumping the clutch and shifting between first and second gear. Every 3 to 4 seconds, continually throughout the entire trip, except for the short distance, we could drive on the 4-lane highway after paying the toll fee.

The 4-lane highway was a welcome sight; too bad it only lasted for 20 minutes of the 4 hour trip!

By the time we reached the US Embassy in Manila, my 100% service-connected PTSD, was in control of my every thought.

Bringing forward that overwhelming nightmarish feeling of impending doom, engulfing my very soul?

This was triggered, from within my scarred memories, emotions left over from a very stressful event!

That event had taken place during October and November of 1985. A year after I proudly completed 9 nonstop months of Basic Military Training, Advanced Individual Training in the US Army Infantryman's Military Occupational Status, Airborne School, Special Forces Qualification Course Pre-Faze, Special Forces Qualification Course First-Phase, which consisted of advanced Hand to Hand Combat qualification, Long range patrolling qualification, Wilderness Survival qualification, and Expert Land Navigation qualification among others. Second Faze Weapons, which was qualification with every military light arm's weapon in use throughout the world at that time, from old, to super modern.

I left the Special Forces Qualification Course after 2nd faze as a temporary termini because my 9 year old daughter had been missing for over 8 months. Because of this, the worrying, I was not able to retain the Russian nomenclature on the Russian weapons, on the last day of 2nd faze. Then and only then, I was asked if I needed time away so that I could find her.

I said "Yes!" Then was told by the school commander, to get a 30 day leave so that I could look for her! I would have to be assigned to a regular unit, and then asked to pick any unit I wanted. I picked the 82nd Airborne just down the street! One month after being assigned to the 505th AIR (Airborne Infantry Regiment), my daughter was found.

I went to see her, and then was informed that the 82nd Airborne would not let me return to SFQC, because I was assigned to them under permanent orders.

Did your eye's just open wide?

They should have!

Two plus Two, equals Four!

Connect the dots…

I had been betrayed!

Try as they might, they were unable to un-train my unconventional warfare abilities!

The tragic event you are about to read, happened while I was serving as a highly qualified Airborne Infantryman assigned as a driver / bodyguard for our Battalion XO, in the 505th AIR (Airborne Infantry Regiment) of the 82nd Airborne, United States Army!

I am still affected by that event to this very day!

I had gone for a scheduled dental appointment at Rhodes Dental Clinic, not far from my unit area of the 505th AIR (Airborne Infantry Regiment). As I waited for my name to be called, I looked through magazines while reading a few articles.

When I heard my name called, I put my magazine down and walked in the direction of the voice calling my name as my spit shined military boots professionally carried me forward.

Following the dental specialists, I was seated into a dental chair as she hung the dental napkin around my neck, onto my chest.

A few minutes later, the dentist arrived to sit down beside me. As he held up the x-rays, between him and the light, he reached for the syringe to give me a shot of Novocain. Next he laid down the x-rays with his left hand, while reaching for a paper napkin.

Grabbing my tongue with his left hand and napkin! He proceeded to move my tongue over to the right side of my mouth. Placed the syringe-needle on the outside of my lower left jaw, inside my mouth, just below my tooth! At the same time, the dental assistant used the dental mirror to pull my left cheek out of the way of the needle!

As he said, "This may pinch a little!" he pushed the needle into my flesh, just under the tooth I was to receive a filling in, injecting the Novocain.

Following the injection, and him lying the syringe down! He stood up saying, "I'll be back in five minutes to see if it's numbing a little!"

Upon his return, he asked if I was feeling any numbness, I stated; I don't feel any change at all....

He then picked up the syringe, injecting me with Novocain in the same spot once again. Walking out of the room he repeated, I will check on you in a few minutes.

A few minutes went by before he came back in to ask if I felt any numbness. I stated that "I felt no numbness at all." Again, for the third time, he injected the area with Novocain, and then left the room saying he would be back in a few minutes.

When he returned, he sat down beside me to say "Okay, let's get to work." At this time I stated that "It is still not numb!"

He looks at me saying "Well, okay. I'll give you another shot, this should be enough." Then again, after giving me the shot, he left the room.

After about 10 minutes he came in the room, and says you should be numb now. I stated: "I feel no numbness at all." He then stated "Well, okay! I'll give you one more but that's it, no more. I can't give you more than five!"

I was beginning to question what the problem might be. I had already been sitting in that chair for over 40 minutes. I had never experienced this. Normally, one or two shots in the area and I was numb for six or seven hours.

He stayed in the room this time, carrying on a conversation with the female dental specialists.

After about five minutes he turned to me saying "Well, let's go ahead and get to work. If it's too much for you to bear, we will just have to stop, then re-schedule you for a future visit. I am running out of time! I need to get finished, so I can be ready for my next appointment."

As he started to drill into my cavity, my hands became vice grips on the arms of the dental chair. It was hurting almost like hell.

He stopped, while looked into my eyes stating: Can you handle it?

I had waited six months for this dental appointment. I knew it would be six months before I could get back with another appointment. That's the way it was, in the 82nd Airborne; your lesser medical needs came last! Instantly, I knew that I was going to sit there bearing the pain, while waiting for the numbness from the Novocain to begin.

That's the way government run health care has always been. 6 months to get an appointment, then take what you can get, because it will take 6 months to get another appointment...

I stated, "Yes, I can handle it!"

And I did bear it, painfully!

As my white knuckles bore the strain of my life and death grip on the dental chair arms, they starved for blood flow! It seemed like forever and a day, but only took 10 minutes to complete the drilling and placement of the filling. I was gratefully relieved once it was finally over. "Need I say more?"

Okay, he said "You're all done!"

Getting up from the chair, I felt the tingle in my hands of the fresh oxygen, brought on by the new freedom of blood flow. Walking forward, I covered the distance back to my unit area.

All that afternoon, through the night, up to the next morning's breakfast, the pain in my jaw never lessened. As I tried to open my mouth to place food onto my tongue, the muscles throughout my face and neck, strained as if they were lifting a heavy weight?

By the night of the second day, I was looking like a hamster or chipmunk, who had stuffed their cheeks full of nuts and berries that they were taking back to their hiding place. Even as if I had a wad of chewing tobacco stuffed into the side of my mouth.

This caused me to return to Rhodes Dental Clinic and walking in on the dentist's as he was performing his duties with a patient.

Looking at my swollen face, he stopped what he was doing. Stood up, asked; "How long have you been like this?"

I painfully stated that it had been increasingly worsening since I left his dental chair two days before.

He stated that I needed penicillin, and to wait right there. He would give me a prescription. While waiting, people began to stare in my direction. As if my face was gross to look at.

Meanwhile a sickening feeling was creeping into my consciousness and physical well-being as my jaw muscles became increasingly tighter.

After receiving a bottle of penicillin with instructions on when and how to take it! I returned to my unit area, walked up the stairs to my third floor room to lie on my bunk bed.

The next morning, I was not able to open my mouth to eat. My lips would open; still I could not spread my jaw to separate my teeth. It was as if I had no control over my cheek, jaw muscles. They had a mind of their own, squeezing my jaws together tighter and tighter as the hours slowly moved along.

Once again, I returned to the dental clinic interrupting the dentist as he was working on a patient. Again, he was shocked to see me. Concerned, he stated that he would write me a profile so that I could have total bed rest. Also, that I should double up on my penicillin!

This was welcome news, as I was becoming weaker because of my inability to open my mouth and consume needed nourishment. Including the sorrow of my left cheeks grossly enormous and painful growth! Which was affecting my physical well-being?

I went straight back to my bunk bed falling asleep.
At 5:30 AM the next morning, I opened my eyes to find my room spinning about me. Attempting to get up, I fell to the floor, to begin pulling my way over to the door. Weakly, I

reached up, turned the door handle, and opened the door, to slowly pull myself along the floor out into the hallway.

Feebly, I inched my way along the floor, toward the next door in the hallway. Barely conscious, I attempted to knock on the door. I could just slightly hear the sound of my knocking. I knew that my friend inside would not hear it and wake up.

I concentrated on breathing, focusing on gathering my strength. Picking myself up on my left elbow! I knocked harder on the door. As strength left me quickly, I found myself unable to stay up on my left elbow. I continued to knock with my right fist while lying on my chest.

As the door opened, all I could see before me were two bare feet. Even so, I repeated the word hospital, hospital, hospital through my clinched teeth. My friend leaned down toward me and rolled me onto my back. He then leaned further forward, looked into my eyes as my voice strained to push past my clamped teeth while I said "Hospital!"

He quickly left, soon returning to help me stand to walk down the stairway and out to his car. As we drove through the streets of Fort Bragg, the fresh cool morning air swirled around inside the car. Refreshing my senses, giving me a little strength!

Once at the hospital emergency room, he pushed the wheelchair up to me, stood behind it, as I climbed in and sat down. Waiting for 45 minutes in the air-conditioned waiting room to see a doctor, helped me as I gathered strength.

Once my name was called my friend pushed me in the wheelchair, down the hallway, into a doctor's office. Once inside, the doctor began to examine me, asking many questions.

After explaining as best I could, the doctor began writing on his clipboard. Stating that he was going to give me a prescription for "Milk of Magnesia," because flushing out my intestines would help with the infection!

I could not believe what he was saying, I became totally frustrated. I pushed the issue of my dental visit at Rhodes Dental Clinic with no use. He was determined to give me "Milk of Magnesia."

Government Run Health Care...

As my friend pushed me in the wheelchair down the hallway, toward the pharmacy! I said to him through my unnaturally clinched teeth, "Stop, just stop! I want to go to the emergency dental wing of the hospital. I am not going to take any "Milk of Magnesia." That Doctor is Stupid!"

Sitting in a wheelchair being pushed around, combined with total frustration over the doctor's diagnosis and prescription of "Milk of Magnesia." Helped me to gather enough strength to stand, walk down the hallways of the hospital, stop at each desk, to ask for directions to the emergency dental section.

Once there, as my friend followed closely behind me, I walked up to the front desk and stated that "I need help!" With one look at my face, the dental specialist behind the counter said that I had -- -- --! For me to take a seat, a doctor would see me immediately.

(I do not remember the technical name she used!)

A few moments later a doctor walked into the waiting room calling my name. As I stood up, my friend stood with me. At the sight of my friend standing up to follow me, the doctor said, "That's okay! He'll be spending a few days here in the hospital. You can go back to your unit."

I turned and tried to say thank you, as my friend raised his hands to stop my attempted verbal "Thank you," to say, "Any time!"

Following the doctor, I walked into a room with a dental chair to sit down. He began to examine inside my mouth and ask questions. He then went on to say; I'm going to admit you into the hospital. Tomorrow morning, we will operate.

I stated: Thank you!

(I was very relieved that finally something was going to be done.)

Then taken to a room and hooked up to an, I-V bag!

The next morning, I was given a glass of orange juice to drink. Along with a clear cup of onion soup, then put into a wheelchair, pushed to the elevator and on down to the hospital emergency dental surgical unit.

After taking x-rays, the doctor examined the x-rays, and then sat down beside me. Meanwhile, his assistant adjusted my I-V, then placed a blue medical blanket on my chest and pinned it around my neck.

The surgeon was a Lieutenant Colonel! A surgeon specializing in emergency dental surgery! His rank far exceeded my own! Funny thing about the military, they put the fear of GOD into you, when it comes to the orders given by an officer. "That should never be, in government run health care...."

As he looked me in the eyes, he stated "Do you drink alcohol?" I answered him with; I occasionally have a drink of Jack Daniels, bourbon mixed with coke. He then said, "Well, this is going to feel like 10 shots of Jack Daniels and Coke!" Then injected into my I-V, a morphine cocktail!

As the morphine entered my arm, I instantly felt total bliss and stated, Hmmmmm!

When I awakened, the blue blanket on my chest was drenching wet with blood and saliva. While the doctor looked into my eyes and says, “It’s all over! You'll be taken back to your room now. I'll check in on you later in the day!"

I stated, though my clinched teeth, "What happened?"

He went on to say that he had cut into my left jaw socket and placed a length of folded rubber. So that the fluids that had filled up my left jaw socket could drain out freely!

(**Note:** I had "Lock Jaw" from bad Novocain, and he thought I had an "Abscess Tooth!")

Later that day, around 3 PM, he came by my hospital bed, and examined inside my mouth. He went on to say that it looks like the drain is going to stay in place. So I'm going to order that you be given 4 million units of penicillin by I-V, every four hours, and 24 hours a day until further notice.

30 minutes later a nurse arrived, to change my I-V, with an equally sized bag of brown fluid. As the fluid ran down the tubes of the I-V, into my arm! The vein in my arm became extremely hot. An intense burning sensation followed through the entire vein, moving throughout my whole body.

The burning sensation running through my veins was so intense that my eyes filled with tears. I lie there in my bed, experiencing horrible misery. From time to time, a nurse came in to give me "Extra Strength Tylenol," with a small sip of water or juice.

I was only able to swallow the Tylenol after forcing it alongside my cheek to the back of my mouth and around the back of my teeth onto the back of my tongue.

Over the next 24 hour period, I received 4 million units of penicillin on exactly 4 hour intervals. During the times that the 4 million unit bag of penicillin had run dry! It was replaced with the clear medical I-V solution until it was time for my next 4 million unit bag of penicillin.

Early the next day, an unconscious patient on a medical bed was pushed into my room and up next to the wall.

A car accident victim!

He had just had a cast taken off of his foot and his broken bone reset, then a new cast put back on it.

About an hour later when his morphine wore off, he began to wake up, moan loudly, painfully excruciating sounds that put the fear of hell into my mind. This was a horrific experience to me for I was lying in pure misery myself. It did not take long before I was pressing on the button which summons a nurse.

Once inside the room, she immediately understood the situation, and then left once again. A few moments later, she and another nurse returned to my room. Pushed the moaning man back out into the hallway.

As time went by, throughout the day I received my 4 million unit bag of penicillin into my I-V, on a regular four hour basis. However, then again, the next day I awoke to find that my room had a second occupant. This time it was a young male soldier who was covered in bandages, throughout his entire body.

After he had awakened, we began to talk and tell each other about the circumstances leading up to our being in a hospital bed in serious condition.

He went on to say that he had gone out with friends and stopped at a bar for a couple of drinks. Now, his friends tell him that the last time they saw him. He was walking outside with a woman that worked inside the bar. Further, he stated that when he was found, he was two blocks from the bar.

Further, just as a train engineer saw him lying face first on the tracks, he had put on the brakes but was not able to stop the train before it ran over him. Also, that after the train had stopped the engineer had gotten out, and found him under the train bleeding profusely.

Further, that his wallet was found 300 feet or so, near the train tracks, from where the train had stopped. Clean of all money! It was back where the train had first hit him. Even that, if he had not been hooked and dragged by the trains' underside! It would have cut him into pieces, killing him.

This poor fellow had deep wide chunks of his head (Flesh and Bone) completely gone! Large portions of the muscles in his arms and legs had been completely ripped away.

He explained that he had been there in the hospital for 4 weeks, and that they needed the room he was in for a new patient. This is why he was moved to my room, for his continued recovery.

Each shift had a different nurse, each with different bedside manners. One was so kind, so understanding, it was as if she felt the pain and misery right along with me. Another was a dictator bent on making me get out of bed every morning at 5:30 AM to make my hospital bed. I disliked this type of disrespect, and voiced my discontent! Which brought in the head nurse, who agreed with me and had that nurse transferred.

This brought in a new nurse that came on shift at four o'clock in the morning. She was a large woman, around 32 years of age, with short brown hair, weighing around 250 pounds.

Because the 4 million units of penicillin, filling my veins and going into the socket of my jaw. It flushed out a dark black cruddy gook. It flushed out my jaw socket by carrying the dark black cruddy gook out through the folded rubber. That was surgically implanted into the socket of my left jaw.

This black gook filled my mouth with the most god-awful taste. Which made me have to sleep lying on my left side? Allowing the gook to flow out of the corner of my mouth, and onto a towel that I laid my head on, when I fell asleep?

Because I was laying on my left side and my left arm was where the, I-V needle entered my vain. I slept with my arm straight out over the side of my hospital bed. Poking out maybe 1 foot!

Nearly seven days into the 4 million units of penicillin every four hours around the clock. I was lightly awakened around 4:30 AM by a slight pressure on my left middle finger.

In my grogginess, it felt as if something was rubbing my finger, as my dreams searched for somehow a way to include this! I awakened to find this overweight nurse, lightly pressing and rubbing her vagina against my finger! Groggily I blinked my eyes, realizing that she did not know I was watching. She was not stopping!

As my head turned a little to the right so that I could look upward, my eyes opened wider and focused on her silhouette in the dim light. She was slowly opening her eyes, and focused her gaze on my face realizing that I was awake. Instantly, her eyes opened very wide, she turned away and immediately left the room.

I was totally disgusted, and looked across the room toward the other bed. I could see that the poor fellow who had been run over by the train was sound asleep. He had no idea what had been happening?

I was not in a good mood, and wanted to voice my opinion. With my right hand, I reached towards the button, which summons, whoever is at the desk.

No one came! So after about 45 minutes, I gathered up my physical strength as I had been weakened further from my illness and daily cup of onion soup. So that I could stand beside my bed! Slowly, I used the rolling I-V stand, to support myself, as I walked out into the hallway.

As I looked left, and then right, I could see no one. The hallway was completely empty. I heard someone talking to my right. Walking in that direction, I was soon at the end of the hallway standing in a doorway.

As I entered, I could see that the room was three times the size of my room and had six or seven hospital beds, each filled with a different man. Sitting up on it, in his bed, a man on the far right motioned for me to come over to him.

As I walked up to him, he stated. I've been pushing the button for the last 45 minutes and nobody is coming. My I-V is empty, and now my blood is running back into the tube.

Hearing this I stated that I had been pushing the button for the last 45 minutes myself, and that maybe he should just cramp up the tube, and keep the blood from flowing back into the I-V bag.

After a moment or two we started getting acquainted, and as I explained, though my clenched teeth, how and why I was in the hospital. He claimed with wide open eyes, that he was in the hospital for the same reason. Even that he had come from Rhodes Dental Clinic.

He went on to claim that three other men in the room. Were there because of the same problem and that one man was scheduled for a 7:00 AM operation? The surgeons were going to open his jaw and scrape out all of the bone marrow.

Because the infection was spreading wildly, and if they could not stop it, it may get into his brain.

Looking at me with wide open eyes he said, "It was bad Novocain, and they're trying to cover it up!"

The next day when I visited him, he stated that the man had died last night, who had had the bone marrow scraped out of his jaw. It was too late; the infection had gotten into his brain and killed him.

I did not like the obese woman who had used my middle finger to masturbate herself while I was sick and incapacitated. Eventually, I slowly forgot about her after she never returned!

Over the next few days, I return to talk to the man in the other room, whenever I could. Then suddenly he was gone. Other patients in the room said that he had been released.

A couple of days later ended my 4 million units of penicillin every four hours around the clock. So I was taken to surgery, where the rubber tube that had been surgically implanted in my left jaw socket was removed.

By this time, it was a total of 2 1/2 weeks, since I had gone to the dentist originally. So you can imagine that, because I had only drunken juice and onion soup including lying in bed 24 hours a day. I had become very weak through this illness. To the extent of becoming totally exhausted after just short walks.

The next morning, the surgeon came by around 9:00 AM to examine me. Then he went on to say that he was going to have some tongue depressors sent by the nurses. And that I was to start with one and force them between my front teeth. Upper and lower and begin to pry my mouth apart.

I do remember feeling a little excited that I was moving into the healing stage. As I hoped that I would soon be eating solid food again.

Once the tongue depressors arrived, a nurse tried to help me by taking a tongue depressor and force it, in-between my teeth. She could barely do it and soon left me to the task.

Slowly, hour by hour, I would push a tongue depressor between my front teeth and try to pry my mouth apart. After a couple of days, I could add a tongue depressor and have 2 tongue depressors one stacked on top of the other forced between my front teeth. As I moved the outer portion of the tongue depressors up or down or tried to twist them in efforts of spreading my jaw apart.

On the third day, the surgeon came by to examine me and had me demonstrate to him, my exercises with the tongue depressors. He seemed concerned that I could only get two after two days. So I explained that each morning. I had had to start over again with one, because my mouth had hardened closed again while sleeping. Further, that it was only until the afternoon of the second day that I could get two in place.

As he left, he stated that he would be by tomorrow, and he wanted to see four tongue depressors stuck between my front teeth. And, that I was to start sleeping all night long with the tongue depressors in-between my teeth, sticking out of my mouth!

As time went by, day by day I could add one more tongue depressors stacked on top of the others in between my front teeth. As you might imagine, when you get to 7, 8, 9, 10, and you have to sleep with 10 tongue depressors, stacked one on top of each other in your mouth, so that your mouth will not tighten back down overnight! It leads to some very horrific nightmares!

By this time, the surgeon was getting aggravated each morning when he came by to see me, and I had only built up to where I could put 10 tongue depressors stacked on top of each other in-between my front teeth.

He was beginning to grind his teeth together when he listened to me, explain the difficulties I was having.

When he left the room that morning, he told me that he was going to return in three or four days, and that he expected me to have 17 tongue depressor stacked one on top of the other in-between my front teeth. And that if I did not, I would not like what he was going to have to do.

This was shocking, as I believed I was doing excellent. I was moving ahead steadily, and could chew a little. Even becoming more tolerant of sleeping with my mouth spread wide open with the tongue depressors jammed between my front teeth!

Three days later, when the surgeon came by in the morning to examine me. He said "Okay, I want to see 17 tongue depressor stuck between your front teeth. Show me!"

I tried and could only get 14 stuck in-between my front teeth.

This angered him and he went off, raising his voice at me saying that, when he came back by tomorrow morning. If I did not have 17 tongue depressors stuck between my front teeth, that he was going to take me down stairs to the dental office and put a mechanism called a "Dead Heart" into my mouth to spread my mouth apart.

Raising his voice one last time he stated; "YOU WILL NOT LIKE IT!

The next morning when he showed up, I could only get 14 and 1/2 tongue depressor stuck in-between the upper and lower front teeth of my mouth. He became instantly inferior-rated and said "ALRIGHT, THAT'S IT, I TOLD YOU WHAT WAS GOING TO HAPPEN. NOW, YOU'RE GOING TO SEE, AND YOU'RE GOING TO BE SORRY ABOUT IT."

I could not believe what I was hearing. I had no comprehension as to why, he was so angry. I don't believe that I could get more than 14 tongue depressors in between my upper and lower front teeth before this whole tragic event ever took place. It just didn't make any sense. I had truly been trying, I mean, I was ready to get up and get out of this hospital, get on with my life. What is this man's problem? Geeeee!

The next morning he came by my room personally, had me get dressed in my duty uniform and into a wheelchair! He wheeled me to the elevator, we then rode downward. He pushed me down the hallway and into the dental clinic, past the front desk into a short hallway and into a room with a dental chair. The entire time with a stern look in his eyes, and on his face!

Then he turned to face me, and with a loud thunderous voice, he exclaimed. "I TOLD YOU WHAT WAS GOING TO HAPPEN. I WARNED YOU! NOW YOU'RE GOING TO SEE WHAT I WAS TALKING ABOUT. YOU ARE GOING TO BE SORRY THAT YOU DID NOT LISTEN TO WHAT I HAD TO SAY. I'M REQUIRED BY REGULATIONS TO HAVE YOUR MOUTH OPEN TO A CERTAIN DISTANCE BEFORE I CAN LEGALLY LET YOU OUT OF THE HOSPITAL. YOU WILL LEAVE THE HOSPITAL TODAY!"

Ok, I was thinking! Let’s get this done, so I can get out of here....

He turned and opened a drawer in the cabinet pulling out a device that looked much like a pair of vice grips or pliers.

However, on the tips they had grooves for your teeth, which pointed outward in opposite directions, maybe 1/4th inch deep. In-between the handles were what looked like slides with little grooves, much like a slide on a Jeep Jack. That work by allowing a small lever to slip in-between the grooves as it is spread wider and wider while locking it into place. So that it could not close.

Nevertheless, with this device of which he called a "Dead Heart" the more he squeezed the handles together, the more the tips of the device spread apart.

He walked over to me and put them into my mouth. Quickly, adjusted it to where my teeth, upper and lower, were positioned exactly on the grooves at the tip of the device.

Then with both hands, he grabbed the handles and squeezes tightly, wrenching my mouth apart as wide as it would go.

My jaw was forced down onto my chest as I weakened from his body weight leaning forward, while he squeezed the device.

With my eyes widely open and filled with tears. He stood back, looked at me and said, "You take it, you take it. I'll be back in 10 minutes, don't you move!"

The first 10 minutes seemed like a long time, I believe he actually waited 20 minutes. Even so, when he came into the room, he walked over to me and seen that my saliva was running down my mouth and chin. A big wet spot was forming on my shirt!

He moved over to another drawer at the cabinet, and pulled out a large towel the size of a beach towel.

Placing it under my chin, he draped it along my body as if covering me with a blanket.

He then grabbed the handles of the device again with both hands, and attempted to spread it apart. Not being able to get it to click into the next, more open position, he climbed up on top of the dental chair, put one knee on each arm of the chair to position himself.

Towering over me, he bent forward and grabbed the device handles with both-hands, squeezing tighter and tighter as he said, “I told you, you were not going to like it!"

With his hands shaking and trembling from his exertion of strength to squeeze the handles of the device! The upper portion, on my front teeth, slipped forward into my mouth, off of my teeth, crashing into the roof portion of my mouth.

The grooved part that was to hold my teeth in place or to hold the device in place on my teeth was now crushing into the roof of my mouth.

I jolted underneath him, squirmed, wiggled and moaned. Again he yelled, "YOU TAKE IT, YOU TAKE IT. I TOLD YOU THAT YOU'RE NOT GOING TO LIKE IT. NOW YOU TAKE IT!"

Then as he could not squeeze any tighter on the handles of the "Dead Heart", his knees slipped off of the right arm of the dental chair and hit me in my ribs knocking the air out of me. This slight change in pressure allowed the device to click into place just a little wider.

As he got down off of the chair, tears were running out of my eyes as I pointed into my mouth, trying to explain to him that the device had slipped off of my teeth, and were crushing into the roof of my mouth.

Surely, he would understand!

I could not speak, but pointed and moaned loudly...

He stated that he did not want to hear any complaints out of me. He had told me, what was going to happen, and what it was going to be like.

He left the room!

I was totally weakened. The 4 weeks while lying in the hospital bed with only onion soup for food had left me unable to fight back any further. All my strength had left me. I lay there with my mind in total shock, thanking that the next time he came into the room; he would understand my tears along with my finger pointing.

Then make the adjustment!

Soon, I was seeing black spots, going in and out of unconsciousness.

Every 10 to 15 minutes for the next three hours he would return again. Each time I would try to explain to him by pointing into my mouth. Up towards the roof of my mouth in hopes that he would just look and see that the device had slipped off my teeth, and was crushing into the flesh at the top of my mouth!

He could see that my saliva had completely drenched the beach towel. Further, the tears in my eyes were beyond what the device was supposed to do. Alternatively, what it was designed to do. Surely, surely he would understand!

Nevertheless, he never did!

Each time he came in, he was angered and inferior-rated even further by my insisting that he follow in the direction of my index finger. Angry, he would climb up on top of the chair, several times having to put his knee into my chest. Squeezing the air out of my lungs as he squeezed the device, one click wider every 10 to 15 minutes!

Many times, I tried to free myself, from the device. I just could not do it...

Toward the end, he could not get it to click any further…

At the end of 3 1/2 hours, I know this, because I was looking at a clock on the wall. I was absolutely exhausted, beyond compare. My body was drained of all energy. I was in and out of faint. I had no fight left in me.

The excruciating pain, as the device was pressed into the flesh at the roof of my mouth, had now overwhelmed my thought process, to where I laid in total mental and physical shock. I no longer tried to resist. And, knew how a wild animal feels when they have had their leg caught in a metal trap with no way to escape.

Again, he wanted to lecture me, on how he had warned me, and how he had told me that I would not like it.

As suddenly as it had begun! He climbed up on me, purposely putting his knee into my chest. Pushing down one last time with his hands, arms and shoulders! He was using so much of his strength his entire upper body trembled and shook! He did this to spread the "Dead Heart" apart one last time so that he could move the little lever, which locks it into place.

Instantly, it closed back together, and then he slipped it out of my mouth.

As he got down off of my chest and my lungs filled with air. My eyes clouded with tears as my mouth slowly tried to relax. He turned to me and said loudly, "I DON'T WANT TO HEAR IT. I TOLD YOU, I DON'T WANT TO HEAR IT! YOU KNEW WHAT WAS GOING TO HAPPEN. I TOLD YOU THAT YOU WERE NOT GOING TO LIKE IT. YOU STOP THAT CRYING! I DON'T WANT TO HEAR IT!

I could not control my crying and continued to point into the roof of my mouth.

With a whimpering voice I said, "I only wanted to tell you, that it had slipped off of my teeth, and was crushing into the top inside of my mouth. I only wanted you to reset it, so that it was correctly on my teeth."

The color in his face went pale, as he reached and placed the "Dead Heart" on the countertop.

As he said, "Oh my God!" he moved over to me and opened my mouth, leaned forward and down. Adjusting his line of sight, until he could look at the roof of my mouth!

Not saying another word, he pulled the beach towel off of my body, which was now completely drenching wet with my saliva and tears from top to bottom.

He then hurled the towel into the wastebasket! Then quietly, calmly said; "You can leave the hospital now!"

I sat in the dental chair for around 20 minutes, as I gathered myself together.

I felt horribly betrayed!

Walking out into the dental office, I looked towards the front desk in pure wonderment as to how someone could sit outside the office surely hearing everything that had gone on inside and not done anything about it! To my amazement, the entire dental office was empty of anyone and everyone but the surgeon and me.

The week before, my friend who had brought me to the hospital had come by to see me and asked what he could do that might help me. I had asked him to bring my car to the hospital parking lot. He did; I had given him the keys to do so.

As I walked outside, I had an idea where to look. I had the keys in my pocket because he had returned them the next day and said; I parked your car out by the trees so that you can find it more easily.

I remember him saying; "Go to the far end of the parking lot on the south side of the hospital." I did and there I found my car.

As I got in and started my car, I looked back towards the hospital, relieved that now finally four weeks later, I was leaving. I never wanted to go back there again, no, not ever.

While driving through Fort Bragg, I drove toward my unit area. Not ready to face anyone with my face still wet from tears, I drove on past, turned right and headed down toward Yakin Avenue and on toward Wendy's Hamburger Restaurant.

I had not eaten solid, real food in a month. I was going to treat myself to a Wendy's Classic Double Bacon Cheeseburger and Fries, with a large Coke!

As I drove up into the Wendy's Hamburger restaurant parking lot, and parked my car. I looked into my re-view mirror. While I wiped my eyes with an old dirty T-shirt that was lying on the front seat!

Then I got out and walked in and on to the bathroom. Under the light, while washing my hands, I looked into the mirror and could see that my eyes were bloodshot and my face was a grayish pale. Bending over I cupped my hands and filled them with water, then splashed the water onto my face. Reaching I grabbed the paper napkins on the wall dispenser and wiped my face dry.

Once out at the front counter and the register. I ordered my Wendy's Double Classic Bacon Cheeseburger with Fries and a Large Coke.

After I paid for and received my tray of food. I walked out into the dining area and sat at an empty table for 4. It was over by the glass windows on the west side. This way I could watch my car as it sat in the parking lot.

As I sat there and reached to pick up my classic cheeseburger! I noticed in my peripheral vision the people sitting in the dining area, were all looking at me, as if I was some dope crazed, drug fiend. Making them wary of my presence!

As I put the cheeseburger toward my mouth and strained to open it wide enough. I was only able to get my mouth open, three quarters the distance necessary to take a full bite of that double cheeseburger.

As my eyes filled with tears my thought process clicked.

Inside my mind, a two-sided conversation was starting. One side of me would ask a question or make a declaration. The other side would decide the answer to the question, even to agree or disagree with the statement. It sounded like, myself inside myself.

I think you call this stewing over the event! As I had no means of venting this over whelming psychological tragedy! It was as if I were multi tasking inside my own mind!

I had lost it, temporary insanity!

It was at that moment, that my 100% service connected PTSD had begun, permanently altering my thought processes.

Well, he had done it! He had broken this Special Forces, trained soldier. It took him 4 weeks, millions of units of penicillin burning their way throughout my veins and 3 ½ hours of torture with a device called a "Dead Heart" to do it!

I would soon be promoted to the rank of a Sergeant E-5. Placed into a leadership position within an airborne infantry squad! Much more experience than my peers, but with a whispering discussion continually affecting my judgment. Always coming forward during periods of extreme stress!

My life continued and compounded this PTSD with experiences in a pre-war, and then in two wars on two different continents, half a world apart. Which you will experience along with me from time to time in this book!

The day I reported back to work at my unit, the battalion XO that I was assigned to drive for, asked me, if "I was given a medical profile when I left the hospital!" I replied "No!" Then tried to explain what had taken place with the "Dead Heart." Looking me in the eye's he said, "Then you're ready to get back to work!" I said, "Yes Sir!"

My thought process was not motivated by fear, but my desire to respond in the correct manner to the problems before me. I would think about every little thing that happened, considering each alternative, but ultimately the answers didn't come from logical analysis. There were just too many variables!

Each time I have claimed this event to the Veterans Administration, they say that my records show that I had an abscess tooth and to drop it!

Tell me, what does that have to do with being tortured with a device called a "Dead Heart," when I had committed no crime? It does not matter if it was an unfortunate event or not.

My claim is not about the tooth, it is about the tragically unfortunate event with the "Dead Heart," which is the root of my PTSD problem. Everything else has grown from that day.

Note to reader: Before I had written this chapter, I had gone through life trying to find the person I use to be, I just couldn't do it! I'm wondering, now, if writing this book will help me....

Funny thing though, now over 20 years later, I still have that tooth, and it has never given me any problems.

Now you have a wide-open perspective, and you are with me...

You realize it is two decades later; we are walking down a sidewalk in Manila Philippines, headed toward the pedestrian crossing, across from the United States Embassy.

As we cross the street and approached the US Embassy, 2 Philippine army trucks and a heavily armored military vehicle, with a Philippine soldier manning a high caliber machine gun, positioned on top, are performing guard duty.

We feel tense, as this brings back deeply ingrained memories of military thugs in the country of Panama, and the dictatorship of "General Noriega!"

Once off the street and on the sidewalk, we turned in the direction of the main entrance to the US Embassy. There we stand in line with about 100 people, all non-US citizens. After a few minutes, we come to a non-US citizen in a security guard uniform, checking IDs. We show him my US passport and are asked by him "Why are you here?" We state that I am here to go to the Veterans Administration office.

He allowed us to pass.

Next, there is a door to enter with another non-US citizen as security guard. There also; we are asked, "Why are you here?" Again, we state my purpose. Again, we are given permission by this non-US citizen to enter the front door. Once inside, we are told by a non-US citizen to step through the metal detector. We do so, and move on through the next door.

There, on the wall, are the names "Veterans Administration and Social Security Administration."

Immediately following the Veterans Administration and Social Security Administration words on the wall are colored lines along the wall leading us in the direction of the appropriate office!

Following the line that leads to the Veterans Administration, takes us down a hallway, a right turns down another hallway takes a slight left turn, and then back to the right again.

Here is a security check with another metal detector manned by a non-US citizen. Once again, we have to show my passport and state my reason for being inside the US Embassy. And once again, we are given permission by a non-US citizen to go through the metal detector.

Once through the metal detector! We follow the same color line on the wall that leads us into a waiting room. Further on, through that waiting room to a turn left and another hallway, this leads us to turn right in another hallway, which ends in another waiting room!

In this waiting room, we have to take a number and wait for our number to be selected. Once our number is selected! We are allowed to go up to a window by a desk. There we speak to the United States Military Veteran's Affair representative, again, a non-US citizen.

Then finally, I am able to turn in my Marriage License.

Inside the same waiting room, is a window to the Social Security Administration of the United States of America?

While waiting for our number to be called, we notice that, the people working on the other side of the Social Security Administration window are all Filipinos.

And to top it off, 2 men and 2 women walked up to the door from the hallway. Knocked on the door, and waited until the door was opened, then they went inside.

Once inside, they stand around laughing and joking. While, 20 people sit in the waiting room along with us. All waiting to be seen by the Social Security Administration of the United States!

What I'm trying to point out here, is, the United States Embassies are manned and maintained by about 90% non-US citizens. This compounded my PTSD…

It was just wrong!

After leaving the US Embassy, I needed money to fund the trip back home. So I stopped at an ATM and withdrew $200. As I put the money into my wallet. The ATM printed out my receipt. Reading the receipt I notice that my balance was not what it should be. In fact, it was much less than what it should be.

After getting on the Internet and checking my bank account online. I found that my Social Security Disability checks, that I had been receiving since 1992 had not been deposited into my bank account. This was a shocking incident as it had never been late before.

I assumed there was some kind of mistake, which would be corrected automatically.

As the days and weeks slowly went by, and the value of the US dollar staidly depleted! It became increasingly more difficult to support "My Hun and my son Bubby!"

What had taken $100 to change into 5,000 pesos, was now taking $150.00. With no hope in sight, the US dollar seemed to be spiraling into a bottomless pit.

This kept my PTSD steadily at the front of my thought processes. Making my day-to-day life more and more difficult to deal with! I was becoming more irritable as each day went by.

Not to mention that I was stewing over the horrid events that had caused my PTSD to begin, years before. Which used up all of my mental energy and left me sleeping most of the day.

Chapter 2

My Little Angel

I found myself nearing the poverty level that so many people here in the Philippines had known all of their life. I had even gone so far as to pay off the $1,500.00 amount which was remaining on "My Hun's" mother's property, and bought several hundred dollars of concrete block. Including, other building materials, over several months, for her mother a new, but small house!

Although it is a modest home, it is the newest house in their neighborhood!

Before I entered their life, she, "My Hun's" mother, along with "My Hun's" sister and her 1-year-old baby girl, were living in a makeshift home. Which, was made from a number of cut down tree limbs, used as a frame and then covered with old rusty sheets of tin? A dirt floor, no running water, with an electric cord stretched from trees to a tree from the neighbor's house.

They had been there for many decades!

They were using an outhouse type shed for their toilet, just like we American's had done 100 and 20 years ago, out in the western frontier.

"My Hun's" living with her aunt had saved her from many years of living in this lower poverty level. All brought on because of her father leaving them for another woman and the babies he had had with her.

Now, 25 or more years later, the father of her sister's baby girl was only coming around once a month with a few pesos. He soon left, saying that his new girl friend was pregnant and has never come back, even unto this very day!

"My Hun's" sister's little baby girl, is now 3 years old, and if she had had to wait for him to bring her food, she would have starved to death long ago.

Yes, that's right! I provide for her, and love her dearly!

Deeply ingrained in my heart of hearts, is an adoption that needs no words on paper. Money will come out of my pocket to pay for her needs, before food goes into my very own mouth!

She is "My Little Angel!" She calls me "Papa!"

On the internet, Wikipedia states: *Angels are usually viewed as emanations of a supreme divine being, sent to do the tasks of that being. Traditions vary as to whether angels have free will or are merely extensions of the Supreme Being's will. While the appearance of angels also varies, many views of angels give them a human shape.*

"My Little Angel" is, "Simply Precious" in my site!

Chapter 3

Serious Need

I had grown up alongside my brother and two sisters while our mother, on the appropriate 1st and 3rd or 2nd and 4th Sunday, took us to the nearest "Primitive Baptist Church" that was having services on that particular Sunday.

Did your eye's just open widely? You see, the "Primitive Baptist Church," holds to the old traditions, from a time in American history when the church's pastor had to travel across long distances to attend the church meeting! And, because there were not enough pastors for each church, the pastors would be alternating between two churches in two different areas.

These old traditions also hold true to the fact that musical instruments were not used, partly because they are for entertaining the congregation. So they have never been brought into the present day "Primitive Baptist Church's."

You also will find that certain individuals with more beautiful voices are not singled out to sing alone or in a group for the entertainment of the congregation. In the "Primitive

Baptist Church," everyone sings together, all at the same time, and is an emotional event. This is part of the camaraderie between members!

I do not speak with animosity toward churches that have and use musical instruments. Nor do I feel bad for those who have beautiful voices and are singled out to sing alone. In fact, I love them! What this statement is about is my deeply heartfelt love for my religious upbringing. It is with my soul that I say, "Surely GOD knew me, before I was ever conceived!"

The pastors that affected me the most were those who would preach straight from the pages of the Bible. At the same time, they would use a personal experience of their own. Adding an eyewitness account to their lessons!

Primitive Baptist, are also famous for the wonderful home-cooked meals after church service! Staying true to the old traditions, each family brings enough food to feed their family and a few visitors. This way, there is always enough food to go around!

So as you might imagine! I'm the type of person, who, quietly but firmly repeats, the Lord's Prayer when I hear the "American National Anthem! To the extent, as if addicted, inside my mind, would say the Lord's Prayer repeatedly! Each time I was given the word that I would be jumping at night. From a perfectly good aircraft, with a parachute that was packed by someone who I had never met, or ever seen, with my own eyes, in the 82nd Airborne, United States Army!

Late one night during the year 1986, while serving in the 82nd Airborne! I had to jump from a C-141 Star Lifter, along with the rest of my unit at around 800 feet. We were practicing military maneuvers so that a new lieutenant could gain experience.

My stick or group was to jump out the left door of the aircraft; this meant that while shuffling towards the rear of the aircraft, I would be turning right to go out the door.

As I shuffled towards the door, and neared it, I was sucked out, as if by a vacuum!

Then as I counted 1000, 2000, 3000, 4000, I realized that I should be feeling the opening shock as my parachute filled with air, but continued to count 5000, and then 6000.

At that instant, my eyes were opened very wide. I could see by the star light, that the entire group of paratroopers who had gone out the door before me, and should be down under me, spotted off into the distance. We're all going straight up at an exceptional speed?

With my eyes widely opened I continued the upward movement of my head and eye's to look up at my parachute. It was not open, it was completely collapsed. I was plummeting towards the earth at a continuously increasing speed!

It was far too late to pull my reserve parachute!

At a speed many times the speed of light! My very DNA knew the calculation, saying from within the depths of my soul. That I was ready to splatter upon the earth and be wiped from the list of living human beings!

This moment traveled many trillion times faster than the speed of light.

Born from the knowledge of insurmountable odds, instantaneously emerged from deep within my very soul, a cry for help beyond cries for help! With the name of The Son of the King of Kings, The Fishermen of Souls, The King of all Angels, The Son of the Living God known to me as The Alpha The Omega, The Beginning The End, The A The Z, The Great "I Am!" had instantaneously built in my lungs. To the extent, my lungs filled with energy so profound that I am not capable of describing it.

Escaping over my tongue and through my lips came a cry for help, that need not be explained. Nor may it be easily explained in mere words, forming itself as a thunderous "JESUS!"

As powerful as any power has ever been, my parachute was instantaneously filled with more air then it was designed to capture. As if the very air itself that left my lungs forming the name "JESUS," filled it beyond its own capacity.

Knowing, I must be very close to the ground as the treetops in the near distance was parallel to my descent. Began, the verbalization of a sentence! As if a broken record repeating itself all while my eyes were clenched tightly closed!

I pleaded out loud in the darkness as fast as I could speak, "Please Jesus, don't let me land hard! Please Jesus, don't let me land hard! Please, please don't let me land hard Jesus, please please don't let me land hard…"

Immediately, my feet touched something soft, something that felt like a pillow. As my weight carried me forward, now my knees were touching the softness. This pillow!

As forward motion and gravity clutched my full body weight! I now pressed against this softness, as my head and shoulders leaned forward. So that I was pressed against the softness, as if I was in the fetal position. With my back toward the heavens! Even, as if I were on my hands and knees, before The King of Kings… With my head bowing in total humbleness and subservient respect!

As my parachute collapsed on the ground around me, and I opened my eyes! I put my hands down before me into the softness; I lifted my eyes, while gratefully saying; "Thank you Jesus! Oh, thank you! Thank you Jesus! Thank you! Thank you, Thank you Jesus, Thank you!"

As my eyes focused, there in the darkness all around me was grass and weeds. In the near distance, trees fully grown, surely by more than 2 decades! There directly under me, shining in the starlight was a white sandy place only large enough for my body to fit into, while curled up.

I had landed in the only soft, sandy spot, within sight! So slowly, as if strong hands were caring me and gently lay me down!

So as you might realize! When I chose to be baptized, I wanted complete submersion! Wanting this, I went to the "Primitive Baptist Church" that I was raised in.

Because there, I knew I was among others who just like me. When they look in the mirror, see the container of a wretched soul. Who would've been wiped from the pages of time long ago? If not for our savior, the "King of Angels, Christ Jesus"!

When I sing "Amazing Grace," tears fill my eyes!

"Amazing Grace, how sweet the sound,
That saved a wretch like me....
I once was lost but now am found,
Was blind, but now, I see...

Renting an apartment and buying new furniture had totally depleted my small savings and left us living from pay check to pay check.

My Hun, offered to go back to work at her aunts small store to help. Even so, I could not stomach the thought that My Hun was going to have to work for less than $3.00 a day, with an unspecified number of hours on any given day. It may be 10, or it may be 15, depending on the stores needs, so she stayed home with me.

My answer to the problem was to stay home more. Stop renting a car each month for any reason other than the necessary trip to Manila and the Office of Immigration each month for my and Bubby's visa.

When the first week of each month came to pass, we had rented a car and made the excruciatingly painful trip to Manila and onto the office of immigration. We then returned home to turn in the rental car!

Every first week of each month, after paying bills, and the trip to Manila, my Veterans Administration Disability Compensation check was so depleted, that we sat home watching TV, or surfing the Internet. With only a once a week or so walk to the movie theater.

As you might imagine, this became increasingly more difficult as time went by, and the value of the US dollar continued to decline, always keeping my PTSD in an aggravated state!

I knew that I must soon find an answer as to why my Social Security Disability Compensation checks were no longer being deposited into my bank account. Somehow someway, I must find out the truth and regain what I had lost.

It had become my main focus in life! I would have my answers!

I had to withdraw money from ATMs, which automatically give pesos by converting the peso to the dollar value of that particular hour. As the days, weeks and months went by, my financial funds diminished totally! Because, the US dollar was daily, becoming worth less and less, against the Filipino peso!

Without, the Social Security Disability Compensation I was quickly becoming a poor man, with "My Hun, My Angel" and my son "Bubby" relying on me to support them!

By September 2007, I was in "Serious Need" of finding out what happened to my Social Security Disability Compensation.

In this frame of mind, I sent an e-mail to President Bush. See copy of the e-mail next:

And please know that the internet service I was paying for was very bad. It worked sometimes, and then sometimes it did not work at all. In fact, I spent more time trying to get online then I spent online. It was driving me crazy!

----Original Message-----
From: [redacted]
To: *president@whitehouse.gov*
Sent: 9/24/2007 12:48:26 A.M. Taipei Standard Time
Subj: Purple Heart Needs Help

Dear Kind Understanding and Honorable President Bush. Have I somehow been magically cured?

Dear Mr. President! Please take a moment of your time and appoint someone to look into this.

My name is [redacted] SSN [redacted]

During your fathers' presidency, I served in the forces which protect our country and our way of life. While doing so, I was prepared to give my life in their defense.

I was wounded in combat, during "Operation Just Cause." During the Gulf War and due to problems beyond my control, I became very sick. After leaving my loved military, I filed for disability compensation and won my case after a 5 year battle.

Over the years that have past, I went to the VA Out-Patient Clinic in [redacted] Florida for medical needs and followed advice from the Mental Health Clinic (Mr. [redacted] = If it causes you stress, stay away from it!).

After returning home from Panama (July 2006) where I introduced my son [redacted] to his Panamanian family members, I returned to the VA Out-Patient Clinic in [redacted] Florida for medical needs. Where I was refused an appointment with the mental health clinic, and was told I had to get a medical doctor to recommend an appointment with them first. This had never happened before as I have 100% service connected PTSD. So I made an appointment with a VA Doctor.

I arrived for the appointment with the Doctor and in his office, I told him about asking for an appointment with mental health, and that they said I would have to get a recommendation from a Doctor. He then looked at me and yelled, saying "I Will Not Give You Any Drugs!" I was very taken back and did not ask again. (Where did he get the idea that I wanted drugs?)

A few days later during another VA Doctors appointment inside VA Out-Patient Clinic in [redacted] Florida, I found that they were using my fathers' medical records to prescribe medication for my heart.

Mr. President Sir, my father had died the year before. This troubled me for weeks, and the day I was to go into the hospital for an operation at the VA Hospital "Bay Pines" Tampa Florida, a woman on the phone that sounded like she was 18 but said she was 50! Told me "not to shave my little groin area, they wanted to do that!", and then laughed!

I turned my back and walked away from this. I did not show up!

A couple of months went by; then I asked my son [redacted], to travel to the Philippines with me. Because I wanted to meet the woman, I had been corresponding with for a year and a half, and that I was worried about traveling that far while leaving him alone. He did and we arrived in Manila Philippines on 6 Feb 2007.

On 20 April 2007, I was married.

On 3 May, we took a trip to Manila and on to the US Embassy. After getting permission from a non US citizen, I was allowed to inter the front gate. After getting permission from a non US citizen, I was allowed to inter the front door. After getting permission from a non US citizen, I was allowed to inter the next door and on to the VA window. There I turned in my marriage license to the VA. After this I walked over to the Social Security office and told them that I had just come from the VA office where I turned in paper work for my marriage and asked if I needed to do the same with their office. They said "No" and I left.

To get back to the small house I had rented and to my son, I went to an ATM to withdraw money for the trip. I noticed that my money was not the amount it should be. I then went to an internet cafe and opened my bank account at Florida [redacted] Bank Online and found that my Social Security Check had not been deposited. And still, as of Sep 2007 it has not been deposited into my bank account. I fear that I will not be able to save up enough money to ever return to the USA.

Dear Mr. President, Kind and Understanding Sir! Please take a moment of your time and appoint someone to look into this problem. And if they find that I have somehow been magically cured and no longer deserve disability compensation, may I have my job back in the US Army?

If there has been a mistake, can my Social Security Disability be turned back on? If there has not been a mistake, where is my disability check going to?

Have I somehow been magically cured? If so, kind Sir, I need proof and the Social Security Administration needs to send it to me. If not, I want the person or persons that may have turned off my disability compensation fired out right and completely. I cannot help but feel that this was done to me deliberately and without remorse.

It's just wrong to leave me stranded in the Philippines without the total disability compensation I have been receiving for well over 10 years. Please see Agent Blackwell FBI, for more information about my disabilities.

And please, kind person appointed by President Bush to look into this, have enough respect to contact me through email at [redacted]. And, let me know what's going on. (If you've done your home work, you know that stress affects me deeply, so please inform me as to what's going on. So I can live without the horrid stress, that's building over this issue.)

In closing, Mr. President, please except my "Thank You for a Job Well Done" during your Presidency and thank you for your respect in advance.

Sincerely,
Purple Heart, [redacted]

By September of 2007, I was getting so very angry about the US government taking such a long time to fix my Social Security Disability Compensation problem that I started emailing "Governor Crist of Florida."

I am from Florida, USA. It is the former governor of Florida "Jeb Bush", who helped me after receiving emails from me, concerning problems I was encountering while in the country of Panama. So why would I not trust the current governor of Florida "Governor Crist" to help now.

To follow are copies of these emails.

You see, while in the city of Panama City, Panama from 2003 until mid 2006. I still received my disability compensation deposited directly into my Florida bank account as always. This meant that I used ATM's to withdraw my money and pay cash for rent, food, and so on.

While doing so, many times the ATM's in Panama City, Panama; would take the money "electronically" out of my bank account in Florida without dispensing any cash to my hand. That's right, the owners or bank owners of the ATM, where electronically receiving my money into their bank. The ATM charge was also taken from my bank account.

Always leaving me without any way to get money for a full 24 hours if I had made a withdraw, of the full amount allotted by my (Debit Card) visa check card issuing company for a given 24 hour period!

The times that this happened, of which I had no money in my pocket prior to my ATM visit, left me in a situation where I could not even make a phone call or take a taxi.
I had to wait until the next day, and walk the many city blocks to the bank branch office of the ATM. And asked to talk to the bank manager!

Each time, the bank manager could not see me and had a secretary handle my complaint.

This had happened many times, and it had become apparent, that the banks were running a scam. Because, they had me fill out a complaint form! Giving them info of time and date, ATM location! And always saying that they would do an investigation, and if it was found that I was owed money, the money would be returned to my account.

In the meantime, I contacted my Florida bank and was informed that, they would have to mail me a form to fill out. I was to return it, to my bank in Florida by mail. Once they received it, an investigation would be done.

I was getting angrier and angrier, as it would take 2 weeks for the form from my bank to get to me, then 2 weeks for it to be returned by mail. Then another few weeks for the investigation to be done! All this before my money would be returned to my bank account.

Also, many American's that I knew in Panama, where having the same problem? Once I walked up to an ATM, and while waiting in line! I watched 1, 2 then 3 Panamanians, get money from the ATM and walk away. Then an American in the line ahead of me, had money taken from his account in the USA electronically. And no cash dispensed to his hand.

It soon became apparent, that the banks in Panama were running a multimillion dollar scam. Where they were electronically, removing $250.00 or so at a time, from USA bank accounts and earning interest on it for the 6 to 8 weeks it took for a complaint to be investigated and the money returned to its rightful owner.

Realizing this, I went online and found the email address of the Governor of Florida. I then sent him a request for help in putting a stop to this scam!

I do not have a copy of it; still it went on something like this:

Dear Governor Bush, please help.
I am a combat wounded veteran of the US Army, living on VA disability compensation, in the City of Panama City, Panama. The banks here are running a scam, where they take money electronically though ATM's from Americans and or vacationing tourist bank accounts and not give them any cash from the ATM. Then draw interest on the money for weeks until they have to give the money back.
If a vacationing tourist with a bank account outside of Panama, goes to an ATM to withdraw money. The money is taken from their account electronically, but no cash is given to them. They then have to go to the bank that owns the ATM and have an investigation done. At the same time, contacting their bank and ask that investigation forms be sent by mail to them.
This causes weeks of time delay and the bank that owns the ATM, is now setting on their money. Example: If 3,000 vacationing tourist goes to an ATM, withdraw $250.00, and the ATM electronically removes the money from their accounts but gives them no cash. That adds up to $750,000.00 for that day. Then while the claim investigations are ongoing for 6 weeks, the bank is earning interest from the world bank of 6%. This adds up to $45,000.00! So you can see, that when this is done, day to day. Over a few weeks or months, what the banks make is in the millions.
Would this not be a great way for the terrorist to fund their activities?
Governor Bush, please assign someone to address this problem.

I received an email reply from the office of Governor Bush that went toward something like this.

Dear Sir,
Governor Bush has received your message and understands it fully. And has asked that I reply on his behalf!
Unfortunately, Panama is out of the governor's jurisdiction, so he cannot address the problem directly.
Your message is being forwarded to the President of the World Bank.
Again, the governor understands your message and thank you for contacting him on this matter.
Sincerely,

The next day, while walking around the block from my apartment! I noticed a white van and men working at the ATM. As I walked up, I watched as a man in a business suit spoke earnestly at a laborer in Spanish, as he opened the ATM with beads of sweat running down his face. I understood enough to know, that the man was not to stop until the ATM was in working order.

Over the next few hours, as I walked around the city. I found ATM after ATM, which was being worked on by crew after crews.

Hahaha, I loved it…

Governor Bush and the President of the World Bank had woken up the banking system of Panama. And the ATM problem did not come back for the rest of the years we were there.

So as you read the emails next, to Governor Crist! Understand that I was looking for the same professional response that I received from Governor Bush.

-----Original Message-----
From: [redacted] *[mailto:* [redacted]*]*
To: Governor Charlie Crist
Subject: Fwd: Purple Heart Needs Help

Dear Sir, I am forwarding this copy of my email to President Bush, because I have received no response from his office. And I so dearly need help in this matter. See copy of email sent to President Bush below.
Please, Kind Governor Crist! Take a moment of your time to appoint someone that will look into this problem and give me some kind of a response.

Thank you sincerely for your respect.

Purple Heart [redacted]
[redacted].

Note: At this point, in the e-mail, I had copied and pasted a copy of my e-mail to President Bush and placed it in this e-mail to Gov. Charlie Crist.
This is a copy of the reply I received from the office of Gov. Crist of Florida. See next:

-----Original Message-----
From: Charlie.Crist@eog.myflorida.com
To: [redacted]
Sent: 10/8/2007 9:52:52 P.M. Taipei Standard Time
Subj: RE: Purple Heart Needs Help

Thank you for contacting Governor Charlie Crist. The Governor appreciates your concerns and asked me to respond on his behalf.

As you may know, this is a federal issue and Governor Crist has limited authority to intervene. Your best source of assistance is your United States Congressional Delegation. You can find contact information for your United States Senators and Representative by visiting www congress org.

The Florida US Senators information is as follows:

The Honorable Mel Martinez
15 East Robinson Street
Landmark Center One, Suite 475
Orlando, Florida 32801
407) 254-2573

The Honorable Bill Nelson
25 East Robinson Street
Landmark Center Two, Suite 410
Orlando, Florida 32801
888) 671-4091

Thank you again for contacting the Governor's office. For information about the Governor's initiatives and to subscribe to his weekly "Notes from the Capitol" newsletter, please visit our Web site at www flgov com.

Sincerely,

Leon Brock
Office of Citizen Services.

I am a registered democrat, and have always been one! During Oct 2007, I was following Senator Obama on the news channels and was impressed with much of what he had to say! Then, when I watched him on the news saying something like: "I'm black you have to vote for me!"

I turned away from him. This type of thing happens in 3rd world countries often. I never thought I would see it in the USA. People were actually listening to him and cheering.

See copy next, of my reply to the office of Gov. Crist, once I had received a reply from his office.

-----Original Message-----
From: [redacted]
To: *Charlie.Crist@eog.myflorida.com*
Sent: Mon, 8 Oct 2007 9:59 pm
Subject: Re: Purple Heart Needs Help

Thank you Leon Brock, for contacting me on this matter! It is the most I have received to date.

[redacted]

A few days later, while checking my e-mail. I noticed a suspicious e-mail in my inbox. It had to do with social security disability, but it had not been sent from the Gov.'s office or the president's office. A few days went by, as I stewed over this and became aggravated. I then forwarded a copy of this suspicious e-mail to President Bush and Gov. Crist. See copy of forwarded e-mails below:

-----Original Message-----
From: [redacted]
To: president@whitehouse.gov
Sent: Tue, 16 Oct 2007 9:58 pm
Subject: Fwd: Get your social security disability today.

Dear Mr. President, I recently emailed your office and the office of Governor Crist (Florida) asking for help finding out if my Social Security Disability check was being stolen. On the 8th of this month, I received a reply from the office of Governor Crist (Florida) saying to contact Congress. On the 14th I received this email, see forwarded copy below.

The only way these people could find out that I needed help on this matter, is if they filtered or read my email to you or Governor Crist (Florida). If so, this is a crime and should be dealt with by the FBI.

If they received the information from someone in your office or the office of Governor Crist (Florida), I take it personally and am insulted.

Sincerely,
Purple Heart [redacted]

-----Original Message-----
From: SocialSecurityDisability765@intwmailrw.com
Reply-to: return-306-7@intwmailrw.com
To: [redacted]
Sent: 10/14/2007 10:23:52 P.M. Taipei Standard Time
Subj: Get your social security disability today.

We found these Social Security Disability sites for you.
Get your social security disability today.
Waiting for social security disability? Look here.
You received this message because you are subscribed to Intelliwerks. To remove yourself from this service and stop receiving email messages from Intelliwerks, go to http://www.intwmailrw.com/us/?306 or go here. To read our privacy policy, go to http://www.intwmailrw.com/pp or go here. Please mail comments about this message to Intelliwerks, 244 Madison Avenue #266, New York, NY 10016.

Reader; I assure you, that I have never subscribed to any Intelliwerks! Knowing this, as each day went by; I became more and more concerned!

See below, a copy of the forwarded message to Gov. Crist of Florida!

-----Original Message-----
From: [redacted]
To: Charlie.Crist@eog.myflorida.com
Sent: Tue, 16 Oct 2007 9:52 pm
Subject: Problem

Dear Governor Crist, I recently emailed your office asking for help finding out if my Social Security Disability check was being stolen. On the 8th of this month, I received a reply saying to contact Congress. On the 14th I received this email, see forwarded copy below.

The only way these people could find out that I needed help on this matter, is if they filtered or read my email to you or President Bush. If so, this is a crime and should be dealt with by your office.

If they received the information from someone in your office or President Bush's office, I take it personally and am insulted.

Sincerely,
Purple Heart [redacted]

-----Original Message-----
From: SocialSecurityDisability765@intwmailrw.com
Reply-to: return-306-7@intwmailrw.com
To: [redacted]
Sent: 10/14/2007 10:23:52 P.M. Taipei Standard Time
Subj: Get your social security disability today.

We found these Social Security Disability sites for you.

Get your social security disability today.
Waiting for social security disability? Look here.

You received this message because you are subscribed to Intelliwerks. To remove yourself from this service and stop receiving email messages from Intelliwerks, go to http://www.intwmailrw.com/us/?306 or go here. To read our privacy policy, go to http://www.intwmailrw.com/pp or go here. Please mail comments about this message to Intelliwerks, 244 Madison Avenue #266, New York, NY 10016.

Next is a copy of an automatic reply from the office of Gov. Crist, see below:

-----Original Message-----
From: Governor Charlie Crist
<Charlie.Crist@eog.myflorida.com>
To: [redacted]
Sent: Thu, 18 Oct 2007 8:38 am
Subject: Automatic response to your email

Thank you for contacting Governor Charlie Crist. The Governor appreciates your taking the time to share your views with him or to seek assistance from his administration. Due to the volume of emails sent to the Governor, there may be a delay in responding to your email. Please know that the Governor's office is making every effort to respond to your inquiry.
In the meantime, you may wish to view our online information database, which provides information on current issues and frequently asked questions. You may visit this source by clicking on Get Answers.
Again, thank you for contacting Governor Crist.

Next is a copy of the reply I received from the office of Gov. Crist of Florida:

-----Original Message-----
From: Governor Charlie Crist
<Charlie.Crist@eog.myflorida.com>
To: [redacted]
Sent: Thu, 18 Oct 2007 3:09 am
Subject: RE: [*]Problem

Governor Crist received your e-mail and asked me to respond on his behalf.
I am sorry about this situation and happy to help if I can.
Please feel free to call me at 850-922-0954.

Sincerely,

Rex T. Newman
Citizen Services

Next is a copy of my reply:

-----Original Message-----
From: [redacted]
To: Charlie.Crist@eog.myflorida.com
Sent: Thu, 18 Oct 2007 8:38 am
Subject: Re: [*]Problem

Thank you, I will try to call soon.
Please understand that I am in the Philippines and will have to use a cell phone. The connection may be bad and can cut off at anytime. Which would be a horrible embarrassment?
Sincerely,
Purple Heart [redacted]

The time in the Philippines is 12 hours ahead of the time in Florida USA. I had tried to call many times, and could not get through. So I did a copy and paste from a form on the VA website of the questions I was having problems finding answers too, and forwarded them to Charlie.Crist@eog.myflorida.com

-----Original Message-----
From: [redacted] *[mailto:* [redacted]*]*
To: *Governor Charlie Crist*
Subject: *Dear Governor Crist!*

Dear Governor Crist!

Sir, I continue to have problems finding answers through the VA website: see copy of Discussion Thread below. (Must be read from bottom to top...)

Further, please re-read copied paragraph from an earlier email to your office in **a bold black type** below: I only ask this of you, so that you may better understand, that I try and try to get help from the VA, and it is always a traumatic event...

Dear Governor Crist! Can you help me get in contact with someone here in the Philippines that will help with this 100% service connected PTSD problem? I need a professional to talk to.

Thanks, Purple Heart [redacted]

Copied paragraph from an earlier email to your office in **a bold black type!**

Over the years that have passed, I went to the VA Out-Patient Clinic in [redacted] Florida for medical needs and followed advice from the Mental Health Clinic (Mr. [redacted] = If it causes you stress, stay away from it!).

After returning home from Panama (July 2006) where I introduced my son [redacted] to his Panamanian family members, I returned to the VA Out-Patient Clinic in [redacted] Florida for medical needs. Where I was refused an appointment with the mental health clinic, and was told I had to get a medical doctor to recommend an appointment with them first. This had never happened before as I have 100% service connected PTSD. So I made an appointment with a VA Doctor.

I arrived for the appointment with the Doctor and in his office, I told him about asking for an appointment with mental health, and that they said I would have to get a recommendation from a Doctor. He then looked at me and yelled, saying "I Will Not Give You Any Drugs!" I was very taken back and did not ask again. (Where did he get the idea that I wanted drugs?)

A few days later during another VA Doctors appointment inside VA Out-Patient Clinic in [redacted] Florida, I found that they were using my fathers' medical records to prescribe medication for my heart.

Mr. President Sir, my father had died the year before. This troubled me for weeks, and the day I was to go into the hospital for an operation at the VA Hospital "Bay Pines" Tampa Florida, a woman on the phone that sounded like she was 18 but said she was 50 told me "not to shave my little groan area, they wanted to do that." And then laughed!

I turned my back and walked away from this. I did not show up.....

Below is copied from the VA website contact form.

Customer ([redacted]) 10/03/2007 07:44 AM
From Purple Heart [redacted] **<---- My *SSN***

I need help with 2 questions.

1.) Over 90 days ago I submitted forms about my marriage in April. Nothing has happened. Why?

2.) My Son [redacted] is 19 now and finishing his High School. He wants to get into collage right away. Who does he talk to about Educational benefits as my dependent?

Thank you for your respect,
[redacted]

Response (316MJ) 10/15/2007 11:25 AM
Thank you for your inquiry. We reviewed your record and found that you are referring to your compensation payment. Please contact them at 1-800-827-1000.

Have a good day

Status: Solved
Customer ([redacted]) 10/16/2007 08:55 AM
Solved? Wrong!

What about question number 2? My Son and I are in the Philippines, do we go to the embassy and ask permission from a non US citizen to get in the front gate, and then ask permission from a non US citizen to get in the front door, and then ask permission from a non US citizen to visit the VA office? Or would it be faster to forward this message to the US Congress? Or to president@whitehouse.gov

Response (316MJ) 10/16/2007 01:01 PM

If you would like to do any of the above things that you stated, you have the right to do so. However if you are not a 100% disable veteran, your son would not be able to qualify for the dependent education assistance. If you are a 100% disable veteran, your son will need to apply for these benefits by completing an application. The application is located on our website at www.gibill.va.gov. The form number is 22-5490.

Customer ([redacted]) 10/16/2007 07:24 PM

I am a 100% service connected disabled veteran (PTSD), and I wish you had done your homework before you attempted to rush and blow me off.
Taking a closer look at the first line of my first contact tread, would have answered that for you.
I will direct my son to form number 22-5490. And in doing so, I hope you figure out, that we (The Vet's) look to you (The VA) for respect. When we do not find it, it feels like a knife in the back and cut's deeply into our souls.....

Further, now that I'm going to be sick all day! Who do I contact here in the Philippines for my medical needs?
Response (316MJ) 10/17/2007 08:30 AM
Mr. [redacted]

This is not an environment to argue back and forth or try to put the other person down or threaten when you are seeking information. If you ever feel that a question has not been answered, all you have to do is re-submit the question. We deal with education benefits only and do not have access to the veteran disability records to know what the percentage of each veteran's disability. We also do not deal or have access to any medical records or contacts of medical records for the veterans. We have to ask questions of you in order to "figure it out" and try to decide what is needed. Hopefully your question has been answer and you are now on the right track. Also remember that everyone deserves respect.

Have a good day

Discussion Thread
Customer ([redacted])
10/17/2007 09:15 PM

Yes, we agree, everyone deserves respect.

Further, I am not arguing or trying to put you down, when I say that "My question has not been answered, and you have not asked any questions to try and "figure it out".

Question not answered: Who do I contact here in the Philippines for my medical needs? i.e. PTSD 100% Service Connected and over active?

If you do not know, please kindly say "I do not know!"

Thank you, Mr. [redacted]

Then I received this reply from the office of Governor Crist of Florida: See below:

-----Original Message-----
From: *Governor Charlie Crist*
Sent: *Wednesday, October 24, 2007 2:47 PM*
To: [redacted]
Subject: *RE: Dear Governor Crist!*

Governor Crist received your e-mail and asked me to respond on his behalf.
To assist you, I forwarded your concerns to the Department of Veterans Affairs for review and response. You should hear from someone in that office.
Thank you for writing and I hope this matter can resolved.
Sincerely,
Rex T. Newman / Citizen Services

I then received this email, see it next:

-----Original Message-----
From: BauerB@FDVA.STATE.FL.US
To: [redacted]
Sent: 10/26/2007 9:08:43 P.M. Taipei Standard Time
Subj: FW: Dear Governor Crist

Your e-mail was sent to me for response. There is a VA Regional Office and a VA Medical Clinic in Manila. They should be able to answer your questions. The address for the

Regional Office is 1131 Roxas Blvd, Manila. The address for the Medical Clinic is 2201 Roxis Blvd the phone number is 632-833-4566. They should be able to provide the treatment you need.

Bob Bauer
Bureau Chief-Field Services
Florida Department Of Veterans' Affairs
Bay Pines VA Medical Center
727-398-6661 ext 4905

From: Bowron, Rupert (Sandy)
Sent: Friday, October 26, 2007 8:24 AM
To: Bauer, Robert (Bob)
Subject: FW: Dear Governor Crist!

From: Schiellerd, Nancy
Sent: Friday, October 26, 2007 8:08 AM
To: Bowron, Rupert (Sandy)
Cc: Daniell, Walter (Earl)
Subject: FW: Dear Governor Crist!
Please handle as appropriate. Thank you.

I then sent this reply to the office of Governor Crist, see below:

From: [redacted] *[mailto:* [redacted]*]*
Sent: *Friday, October 26, 2007 10:36 AM*
To: *Bauer, Robert (Bob)*
Subject: *Re: FW: Dear Governor Crist!*

Thank you, I will contact them.
May I contact you though this email address (BauerB@FDVA.STATE.FL.US), if I have any more problems? [redacted]

I received this email in response, see next:

-----Original Message-----
From: BauerB@FDVA.STATE.FL.US
To: [redacted]
Sent: 10/26/2007 10:46:21 P.M. Taipei Standard Time
Subj: RE: FW: Dear Governor Crist!

Yes

Bob Bauer
Bureau Chief-Field Services
Florida Department Of Veterans' Affairs
Bay Pines VA Medical Center
727-398-6661 ext 4905

Then I sent a copy to Governor Crist, see below:

-----Original Message-----
From: [redacted]
To: Charlie.Crist@eog.myflorida.com
Sent: Mon, 29 Oct 2007 4:56 pm
Subject: Fwd: FW: Dear Governor Crist!

Hi Rex T. Newman! Out of respect for you and Governor Crist! I am forwarding this email sent to Bob Bauer at the VA in Bay Pines, so that you may remain informed...

Thanks,
[redacted]
--

-----Original Message-----
From: [redacted]
To: BauerB@FDVA.STATE.FL.US
Sent: Mon, 29 Oct 2007 4:53 pm
Subject: Re: FW: Dear Governor Crist!

Hi Bob Bauer,
I was able to call the number you gave me in Manila and after a few tries, even to get though. After questioning, I was forward to extension 3203, where a polite young lady asked for my service number. After 1 minute she confirmed my service number and service connected disabilities adding up to and over 100% combined.
But was not able to give me an appointment for the next 3 months! She went on to say that she has sent out emails to Doctors hoping that they will come on board. She then asked

for my contact number, which I gave her. And after explaining, "that if a Doctor or Doctors replied to her emails," she might be able to get an appointment for me sooner, maybe even in Dec. of this year. And asked if she could contact me next week!

I have to assume that they are understaffed and underfunded. Another click or 2 against my stress level!

I took it further and asked who my wife might see for any medical needs she might have in the future! She informed me that a Champ V form would have to be filled out. Question, do you know the exact page I might find the Champ V Form on the VA website? Or will it do any good because of the paper work for my marriage being in the computer but not finished, as I was told by the VA Rep at 412-395-6272?

(I only today, after this phone call, informed my wife that she is covered as my dependent though the Champ V, for her medical needs!)

Because it has been sense the first week of May 2007 that I turned in the form concerning my marriage paper work at the VA office in the US Embassy Manila. And it has not been finalized in the VA records system. Are they "the VA records system," understaffed also?

Thanks,
Purple Heart [redacted]
[redacted]

On the 26th of Oct 2007 I had received an email from the Office of Congressman Tim Mahoney, see copy that follows:

-----Original Message-----
From: john.mac@mail.house.gov
To: [redacted]
Sent: 10/26/2007 5:02:30 A.M. Taipei Standard Time
Subj: Your Social Security Issue (CONGRESSIONAL INQUIRY)

Mr. [redacted]- I am in receipt of your letter addressed to the President. It has been forwarded to Congressman Tim Mahoney's office. I have sent a letter to the Social Security Administration on your behalf and will update you accordingly. I understand your letter and your main concern; I just have a question pertaining to the issues you alluded to. Which months have you not received Social Security payments for and was this money for retirement or disability? As soon as I receive a reply to the Congressman's letter, I will let you know. In the mean time if you should need any additional assistance please let me know.

Sincerely,
John Mac
John Mac
Constituent Liaison
Office Of Congressman Tim Mahoney
9 SE Osceola St
Stuart, FL 34994-2125
772-878-3181
772-871-0651fax
John.mac@mail.house.gov
Visit www mahoney house gov, and sign up for Congressman Mahoney's e-Newsletter
---

I then sent this reply to John Mac after he had contacted me through email, see copy below:

From: [redacted] *[mailto:* [redacted]*]*
Sent: *Thursday, October 26, 2007 11:41 PM*
To: *Mac, John*
Subject: *Re: Your Social Security Issue (CONGRESSIONAL INQUIRY)*

Thank you so very much for contacting me. Worry had engulfed me so badly; I had given up hope of any help from my Gov.

It was May of this year (2007), which started the months of no **"Social Security Disability Compensation"** deposits into my bank account at Florida Community Bank in [redacted] Florida "My Home". Not knowing, I assumed it was a mistake and would resume the next month and be caught up. That has not happened... (I never dreamed I would be stuck here.) Where I was staying at that time "{[redacted] Valley - [redacted]- Philippines}", there was no phone, no ATM, no Internet.

And I spent my days taking a bus to the nearest city "Olongapo", and on to an Internet cafe. Trying to find a way to contact the Social Security Administration in hope of some answers! At their website I found out, they would only give answers after they made contact through my home phone number in Florida.

And that was impossible, which drove me to rent an apartment here in Olongapo, as I needed to stop the daily trips, looking for a working ATM, as they do not work for 24-7 here. Which drove me further into dept., as I had to put stove, refrigerator, TV and furniture into this place? So we can live, until I can save up enough money to get us (my son and I) back to the Good Ole US of A.

To get my wife to the USA, I understand it will take as much as 2 years for the visa to go through. So I have put that on hold and have been running around getting paper work for a 'permanent visa" for my son and me. Because I understand that we will be deported on 6 Feb 2008 without it. Which drives my stress level through the roof?

"The nightly nightmares of impending doom have caused dark rings to form under my eyes." And even the slightest thing seems to push my blood pressure up. Now that I have heard from you, there is a little light at the end of the tunnel! At least, I might find out if my **"Social Security Disability Compensation"** is being stolen or just cut off. If it is being stolen, please catch them and put them in jail! If it has just been cut off, somehow, I need to replace that money. As I have lost over $1500.00 per month in disability compensation sense 2002. This last loss of $900 per month May 2007) has broken the camels' back. My need to file bankruptcy is years past due...

Rex T. Newman, Citizen Services at Charlie.Crist@eog.myflorida.com, who gave me your contact info. He has contacted the VA on my behalf, in hopes that they can find a Professional Doctor, I can talk to here in the Philippines. It has been 3 days and still no contact from them. It took them over 6 months (in 2002) to send a copy of my medical records "Judge Cary wanted," to prove my complete disability. They arrived one week after the trial. Which was too late, and Judge Cary "[redacted] Florida" judged against me? And took my home from me! So with that in mind, I do not look for the VA to help or contact me any time soon.... My son and I ended up living in a barn next to my Dads place for over a year.

Until I gave up and moved my son and I to Panama where he was born, and I first met the Devil. I needed to confront this Demon, who had tormented me sense 1987. And it did help until March 2007, when - while wide awake, the face of a neighbor's child appeared in front of me and stopped my breathing for a moment. He went from a happy and playing boy of 7, to dead 2 weeks later (after his face appeared to me) of cancer.

Just last week, Thursday 18 Oct., while my wife was planning a trip by bus to Manila! I was taking a nap. A dream started, where I was walking with a family member who looked behind us and said, "Why does that man have a gun pointed at [redacted] (my wife)? I ran to a door, went in, then through another door. Where I found a man in a medium to light colored T-shirt, brown belt and jeans! Short graying hair, about 5'10"! Standing behind my wife with a gun pointed to her head. She was wearing jeans and her favorite shirt. I pulled the man from the back of his head and through him to the ground. The dream ended.

When I had a chance, I told my wife about the dream.

The next day, Friday 19 Oct. when my wife was leaving for Manila, I stopped her because she was wearing the same jeans and favorite shirt that she had been wearing in the dream. I refused to let her go until she changed.

She then left!

An hour later she came home and said that she canceled the trip because her paper work was wrong.

Later that same day, the news channel was filled with the story about the bomb blast that killed many people in Manila. 1 block from the building my wife was going to be at. She then told me that she and her cousin were talking about going to that mall if they had time during their trip. Thank God she canceled and did not go. Now it is Friday again, and she has left for Manila. I do not for see any problems!

A few months ago, I had a dream where I was looking at my son as he lies dead in a coffin. I keep praying and rebuking that dream to hell from which it came!

These dreams that come true have been plaguing me sense I was lying in a fox hole in Iraq, during the 1990/1991 Gulf War. On or about the first week of Dec. 1990!

I do not believe I deserve the disrespect; I receive from the employee's of the Veteran's Administration and Social Security Administration. Nor do I believe I should be forgotten by my Government.

I believe that my Government should remember, that when needed and called a pone to defend her, I did not falter, nor did I hesitate to stand up and jump from a C-141 Star Lifter, in the middle of the night while enemy bullets filled the sky around me, so close as to go through my parachute. And after being wounded in one war (Operation Just Cause), I continued to serve from day 1, into and through to the completion of the next war (Gulf War).

I was called at my home on 7 Aug. 1990 for deployment to the Gulf War. I returned home to my hell on March 15, 1991. By May 1991, the Devil was talking to me face to face in the front yard of my home outside of FT. Bragg NC. And after I said, "Why don't you just kill me?" It said, "I don't have the power to kill you, only to drag you through Hell. Look at you; I have drained you like a vampire."

By Oct. 1991, I was too sick to do my job as a member of the 82nd Airborne. And a doctor helped to push me out of the US Army.

If the Social Security Administration has found that I have somehow been magically cured! I plead that the US Government find a place for me in the US Army. Preferably somewhere close to Osama Ben Laden. So that I might end his time on earth!

In my hour of need, I pray that my Government will stand up and fight to the bitter end, in defense of me. Amen, Amen and Amen...

Sorry about that, it's just that I am so stressed up that I needed to vent the pressure a little!

[redacted],

Purple Heart [redacted]

What you have just read, is full blown "100 % Service Connected PTSD," commonly known as "Post Traumatic Stress Disorder." I call it "Totally pissed off, because I was pushed and pushed, until I was on the floor."

-----Original Message-----
From: Mac, John <john.mac@mail.house.gov>
To: [redacted]
Sent: Wed, 31 Oct 2007 4:59 am
Subject: RE: Your Social Security Issue (CONGRESSIONAL INQUIRY)

Mr. [redacted] after speaking to the Social Security Administration on your behalf, they forwarded information to me, which are contained in the attachment. Your benefits had been terminated by the Social Security Administration in February of this year. This was due to the fact that the Miami filed office had been trying to correspond with you to submit a new up- to- date medical evaluation. This evaluation is required by the Administration to substantiate your benefits. B/c you had failed to do this and had not responded to Social Security your benefits had been terminated.

I had also inquired with the Veterans Administration and they have a pending claim for a spouse on your record and that is currently in the review process. No determination has been made to date.

If you should need additional assistance this office can try to assist you, however your Social Security Issue cannot be appealed due to a statute of limitations which has already ran out. You had 60 days from the notice of the enclosed documents to contact them. That date passed on September 14, 2007.

Sincerely,
John Mac / Office of Congressman Tim Mahoney

I could not believe this man! He had obviously only scanned as he read. He had not looked into this very far at all. And was jumping to conclusions, without all the facts! There's no way in my mind that the statute of limitations could have run out. Certainly not when I had just now gotten the message, and I had gotten it from him in an email attachment. This was a photocopy of a letter from the Social Security Administration, sent to me, of which I had never received. Stating that they had determined, that I was no longer disabled, and that I owed them money from being over paid for many months?

I was mad as HELL, and I was not going to take this disrespect anymore!

See my reply next:

-----Original Message-----
From: [redacted]
To: john.mac@mail.house.gov
Sent: Wed, 31 Oct 2007 2:58 pm
Subject: Re: Your Social Security Issue (CONGRESSIONAL INQUIRY)

John Mac, I never thought I would receive disrespect from you.

Note: See also paragraph 13, sentence 4 [As this is on the order of misconduct liking to that which happened during "Hurricane Katrina."]

Today is the first day I have received any notice of this kind from the Social Security Administration. You know that! And it came to me though you at the office of **Congressman Tim Mahoney. Of which there is a record of!**

Sir! You can clearly see that the address these letters from the "Social Security Administration", were sent to an APO address that forwards to the country of Panama. I closed that account in Panama June 2006. If mail was not returned to sender by them, they should be dealt with for it, not me! Emigration records, i.e., American Airline's passenger list and passport stamped at the Miami airport during the first week of July 2006, clearly show that I was no longer in the country of Panama.

Further, the paper work I filled out for emigration when re-entering the USA during the first week of July 2006, shows [redacted] Florida as my destination. Even further, email correspondence to then **Governor Bush,** proves I was in [redacted] Florida and was being stocked by local law enforcement officers. Of which the office of Governor Bush helped to put a stop too!

While reading the letter it is obvious that disrespect is intended. (Look at the second sentence of the first paragraph of your email. "Your benefits had been terminated by the Social Security Administration in February of this year.") I arrived here in the Philippines 6 February of this year.

(You know that!) How could I have been in 2 places at the same time? Further, dew the fact that your office knows how long I have been trying to find out the truth. {Not standing up and speaking out on my behalf by your office will further the injury from this disrespect. And cause these injury's by the Social Security Administration to go unchanged and unhealed.}

Also: Why is it that the Veterans Administration does not update the Social Security Administration with my medical records? One Fax would do it! They do have in their records that Medicare part, A and B, is my medical insurance. (I know this because I updated my insurance records with them.) The VA knows the Doctor's, I have seen in Panama and what medications I was taking. And so does President Bush, as they had contacted him personally by letter because of nonpayment. The VA paid the bill, up to the point where misconduct by a doctor caused me to contact the VA about it!

Am I expected to go to a doctor from the VA and take medication from him or her, then on the same day go to a doctor from Social Security and take medication from him or her? So much for the "Persons with Disabilities Protection Act," and so much for the US government not having to double pay for the same treatment… Gee, won't the doctors love this; they can beat the system and double charge the US government for the same treatment and get away with it by law.

To think that I get jerked around from one Government Administration (VA) to another Government Administration (Social Security) is sickening in itself. So much for the "Reduction of Paper Work Act"!

Look at this from your email: (I had also inquired with the Veterans Administration, and they have a pending claim for a spouse on your record and that is currently in the review process. No determination has been made to date.) That paper work was turned in during the first week of May this year, at the US Embassy here in the Philippines.

How long does it take? Surely, there is not more paper work to be filled out! And if so, will I ever be notified? They're not still trying to use my father's records as mine are they?

I have forwarded a copy of my last email to the VA, to the office of Governor Crist. That proves I tried to get help for this PTSD problem but cannot get an appointment for 3 months. And that they, the VA here in the Philippines are looking for a Doctor to treat me, as we speak.

Are the Social Security Personnel under the same protection as "Black Water?" When are they going to be held responsible for this wrong doing? One phone call by them, to the VA, would have cleared this up, months ago. So this obviously is not something they want to clear up, and that they ether, get a kick out of being disrespectful to Veterans, or they are paid by how many Disability Compensation Claims, they cutoff daily. Sounds like an inter Government "Push Me, I Push You. Let's make them do more paper work!"

And how could I possibly pay any money back to them? They know I'm under doctor's orders not to work! Or do they have their own little dictatorship, and make their own rules?

Am I to believe that they will garnish my VA Disability Compensation to collect this amount, they claim I owe them? Or is this just their way of pushing my "100% Service Connected PTSD" into doing there will? Wrong answer, this could lead to my suing and winning large amounts. It's not funny, nor should it be the job of a person or persons working in a Social Security Office, to over ride the advice of a doctor. (i.e. "Mr. [redacted], VA Out Patient Clinic [redacted] Florida., **"Stop working! If it causes you stress, get away from it!")**

How can I get away from stress that so greatly affects my PTSD condition, when members of the Social Security Administration and members of the Veterans Administration, seem to enjoy putting stress directly on me? And the Office of Congressman Tim Mahoney seems not to care at all! Because with one phone call could end this horrific and or misconduct on the behalf of the Social Security Administration! And have my disability compensation turned back on. Reducing any further paper work or harm to this 100% disabled Veteran.

Further: I was absolutely refused any "Vocational Rehabilitation" by the "Veterans Administration" after I applied and tested for it. So how can I possibly get a job other than that of which I was trained (Airborne Infantry, Guerrilla Warfare, Nuclear- Biological and Chemical Warfare "NBC".)?

Am I to hire a Lawyer or Lawyer's to file charges against the Social Security Administration for harassment and pay them with my Disability Compensation? Or, are they "The Social Security Administration Personnel," protected like "Black Water?" That is just what this is.. Even to hire a lawyer to fight for me against the Social Security Administration, would have to be paid with Social Security Disability Compensation and or Veterans Administration Disability Compensation. As it is all I have.

Imagine having to fight one Government Agency with Disability Compensation from another Government Agency.

That's the US Government paying a lawyer to fight itself! Is there to be a "Congressional Investigation" into misconduct by members of the Social Security Administration? To wit: Stocking and Harassment by members of the "Social Security Administration" against a 100% disabled Veteran? Surely, they got a kick out of this, and know they are protected for doing same. Maybe I should be talking to the "Order of the Purple Heart!" Or "Anderson Cooper 360."

As this is on the level of "Government Misconduct" liking to that which happened during, "Hurricane Katrina"! How many other Veterans or citizens will come forward with similar claims?

Dear Congressman Tim Mahoney, can you not see that I only today received this information and therefore, this entire situation is unjustifiable. And that the only way to make it right is for your office to order, that the Social Security Administration turns back on my Disability Compensation until a full investigation can prove my continued disability of which the Veterans Administration has archived as "Absolutely, Totally and Completely, Permanently 100% Disabled? And that any further contact from the Social Security Administration concerning my continued Disability should be directed to the Veterans Administration. And in not doing so, causes an unjustifiable and unneeded investigation by the US Congress. **(That is the only reason you could care less, isn't it?)**

Because of the stress, this wrong doing against me by the Social Security Administration, and the, "could care less attitude by my congressman." Am I also to open a whole new can of beans and give the name of the officer appointed over me during "Operation Just Cause," who gave out all of his equipment to soldiers under him to be carried by them into combat, so he could fill his backpack with a pillow. So as to, keep it light enough to run like a rabbit when confronted by enemy soldiers. Filling his backpack with a pillow also gave it the look of a fully filled and heavy backpack. Warding off comments about it and his cowering activities in the face of an armed enemy!

And when I confronted him about it, he said "What do you think, I'm stupid! I have to run!"

Within 30 minutes, I was shot by a sniper from his roof top position. While acting as a volunteer guide during a rescue convoy to other unit members, hours later given morphine and sent back to the US.

And that once back at Fort Bragg and confronting this Officer again, I was removed from my job and put under a First Sergeant with orders to make my life hell. Which, he did!

And a year later while in the Gulf War, I received a simultaneous chop to the throat and a sweep to my feet from a fellow soldier, I had found hiding when it was his turn to be on guard duty. Which sent me flying through the air to land hard on my back? Then threatened with my life if I did not shut up!

Is it true that at this point in time, I find myself, defending myself against my own government? When I have done no wrong, and have pled for help on this matter for weeks and months!

You wrote: If you should need additional assistance this office can try to assist you, however your Social Security Issue cannot be appealed due to a statute of limitations which has already ran out. You had 60 days from the notice of the enclosed documents to contact them. That date passed on September 14, 2007.

Because, I have only received these letters from the Social Security Administration today, 31 Oct 2007 through your office! Any previous 60 day limit is void by law. And I have 59 days left to me by law. **I firmly request, that Congressman Tim Mahoney** appoints a person or persons with the authority to invoke the Social Security Administration into reinstating my Disability Compensation until the VA Office here in the Philippines can find a Doctor to properly evaluate my continued disabilities and report them to the Veterans Administration, which in turn can report the same findings to the Social Security Administration. And that the Social Security Administration be commanded to respect the decisions of the Veterans Administration findings on my disabilities. And, to act accordingly with the "Reduction Of Paper Work Act!" And the "Citizens With Disabilities Protection Act."

Thank you Sir, for your respect, in advance! For surely, this matter warrant's action by your office!

Purple Heart [redacted]

That was the last contact I had with john.mac@mail.house.gov. He would no longer answer any emails from me. I could not even get any emails to leave my account when I was sending to john.mac@mail.house.gov. I'm sure they were blocking me somehow. So I sent it to Governor Crist. See copy next:

-----Original Message-----
From: [redacted] *[mailto:* [redacted]*]*
To: *Governor Charlie Crist*
Subject: *Fwd: FW: Dear Governor Crist!*

Sir, somehow my email to john.mac@mail.house.gov is being blocked. I have sent it twice. Once yesterday and once today! Still it is not in the sent box of my email account like my emails that have been sent to you.
I do need you to forward them to john.mac@mail.house.gov and the office of **Congressman Tim Mahoney,** as they are being blocked and not getting through.

Please reply!
Purple Heart [redacted]
[redacted]

Reader, as you can see, that email had the email address to john.mac@mail.house.gov inside the first sentence... What I received in the reply email was shocking…

I received this reply, see copy below:

-----Original Message-----
From: Charlie.Crist@eog.myflorida.com
To: [redacted]
Sent: 11/2/2007 2:34:39 A.M. Taipei Standard Time
Subj: RE: FW: Dear Governor Crist!

Dear Sir:

I don't have Congressman Mahoney's e-mail address. Please feel free to contact me at 850-922-0954 and I will assist you in any way I can.

Sincerely,
Rex T. Newman
Citizen Services

I could not believe it!

Right in front of my eyes, he was lying to me. I knew this, because the email address to Congressman Tim Mahoney's office was right there in the email, I had just sent to him.

So I called him, and while explaining everything all over again, he put me on hold and then came back with a 3rd party call with the Orlando office of Congressman Mel Martinez on the other end.

Again, I had to explain everything all over again; then I was told to fax a signed copy of the Freedom of Information Act release form to them after I printed one off of the internet.

Once I had the form ready, I could not find a working fax machine that could send to an international fax number. So I sent it by FedEx, and because I did not have their email address, I sent word of the FedEx by email to Rex T. Newman, in the office of Governor Crist, see copy to follow:

-----Original Message-----
From: [redacted] *[mailto: [redacted]]*
To: *Governor Charlie Crist*
Subject: *Re: FW: Dear Governor Crist!*

Dear Sir,
I was not able to find a fax to send the form and info needed by Senator Martinez's Orlando Office. So I sent them by FedEx on 6 Nov, 2007.
Tracking number [redacted]
I was told that it should arrive on the 8th of Nov 2007.
Thank you for your help.
[redacted]

This is a copy of thc email I received in reply, see next:

-----Original Message-----
From: Governor Charlie Crist
Charlie.Crist@eog.myflorida.com>
To: [redacted]
Sent: Thu, 8 Nov 2007 12:38 am
Subject: RE: FW: Dear Governor Crist!

You are welcome. Let me know if there is anything more I can do.

Sincerely, Rex T. Newman / Citizen Services

It had been 3 months sense I had been able to make a payment to my internet provider. The next day, my internet account was turned off, because of nonpayment.

My ability to correspond on this matter had now been hit with a serious blow. And, my PTSD condition was hampering my ability to cop.

Reader; You do not have an accrete perspective! Because, at this point, you have questions as to what I was referring to when I mentioned in my reply emails to **Congressman Tim Mahoney,** at john.mac@mail.house.gov about:

1.) The officer appointed over me during "Operation Just Cause," who gave out all of his equipment to soldiers under him to be carried by them into combat, so he could fill up his backpack with a pillow.

2.) Dreams that come true.

3.) Chop to the throat by a fellow soldier.

4.) Met and talked to the devil.

So read on while you gain a more in-depth perspective:

Chapter 4

My Soul

During 1988 I watched as my little "Bubby" was born at Gorgas, US Army Hospital in the Panama Canal Zone. Three weeks later; he was dying in the same hospital of his birth.

Days later, his Doctor walked up to me and said, "I'm going to take one more spinal tap and a blood sample; they had already taken many! From your baby, so that I can send it to Walter Reed! If they do not know what the problem is, I give your son three days to live."

It took 2 nurses and the doctor to hold "My Little Bubby" down, while they drew fluid from his spine. This put tears into the eyes of his mother and me! It's a horrible thing to see a 3 week old baby fighting so hard, that 2 nurses cannot hold his arms and legs still. His scream was that of; the knowledge of pure betrayal!

I immediately went off by myself and began to pray for my little "Bubby;" it was long past time! I then called home to Florida and informed my family; I needed their prayers also!

Three hours later a nurse walked up to me holding a large medical dictionary. She said, "We have received an emergency message from Walter Reed concerning your son. They said he has "Listeria" and to treat it with "Gamma Goblin."

I did not know what "Listeria" was and asked! The nurse replied, "We didn't know what it was either and had to look it up!" She then opened the medical dictionary and read to me the description of "Listeria.

On the Internet, Wikipedia describes Listeria as follows: Listeria monocytogenes is a bacterium commonly found in soil, stream water, sewage, plants, and food.[1] Each bacterium is gram-positive and rod-shaped. Listeria are known to be the bacteria responsible for listeriosis, a rare but lethal food-borne infection that has a devastating case fatality rate of 25%[2] (Salmonella, in comparison, has a less than 1% mortality rate[3]). They are incredibly hardy and able to grow in temperatures ranging from 4°C (39°F), the temperature of a refrigerator, to 37°C (99°F), the body's internal temperature [1]. Furthermore, listerosis's deadliness can be partially attributed to the infection's ability to spread to the nervous system and cause meningitis.[1] Finally, Listeria has a particularly high occurrence rate in newborns because of its ability to infect the fetus by penetrating the endothelial layer of the placenta.[2]

June 1989, I had spent 2 ½ years, of my three-year re-enlistment for Fort Clayton Panama, at Fort Kobbe, in the Panama Canal Zone; another betrayal. I had reenlisted for Fort Clayton and a non-airborne job! Then once in Panama and at

Fort Clayton, was told that if I did not take an airborne slot; I would never be promoted again!

Noriega and his military guns had systematically kidnapped American soldiers and/or American military dependents to the extent that the US government began sending all US military dependents in Panama back to the United States.

This is how my "Bubby" and his mother entered the United States!

Because my "Bubby's" mother did not speak English, my "Bubby" was nine months old, and this was the first time out of her country. It created a hardship, and I was reassigned to the 82nd Airborne, United States Army.

I was given a job as Air Items NCO, under a First Lieutenant in the S-4 section of a battalion in the 325th (AIR) Airborne Infantry Regiment. While, still harboring ill feelings, toward other NCOs, back at my unit in Panama, for their constant bombardment of disrespectful comments about my biracial child!

Such as "How is your Hooch Baby? Is he still alive?" To include bad counseling statements from my squad leader, for going to the medics when I had a bad case of diarrhea!

This was later found to be a parasite in my blood that was keeping me from properly digesting my food!

The many months of patrolling the jungles around "Howard Air Force Base" in Panama, left the Officers in charge very stupid. As they had everyone sleep in the same spot that 400 other paratroopers had been sleeping in for 3 months, with no urinals!

The entire area was saturated with urine and feces as we slept. The urine and feces had built up over many weeks, and when we pointed it out, were ordered to sleep in it anyway!

Hence, my bad case of diarrhea was born, and after 6 months of going to the medics about it, it was found to be a parasite in my blood that was keeping me from properly digesting my food; it was not like regular diarrhea, there were no fevers or intestinal cramps! It was all water!

Once I raised hell about being given a bad counseling statement for it; because I kept returning to the medics each morning before my daily duties began. I was given permission to enter the hospital for test! Then given a very small pink pill to kill the parasite in my blood!

You might think that doing my job every day of the 6 months I had diarrhea, and not missing a moment of required training or live mission's; It would have warded off any bad counseling statements for going to the medics often. Nevertheless, it didn't; my biracial marriage and biracial child brought out the worst behavior in everyone around me!

Once the word of my biracial marriage and biracial child got around, animosity had a way of totally engulfing the minds of the people I worked with!

Dec 1989, I was called at home, and given our call out notification. "Operation Just Cause" had begun!

Once at my unit area I was informed that I would be filling in for the S-4 RTO (Radio Transmission Operator) an E-4, because I had just returned from a tour of duty in Panama and knew the area well. To include speaking a little Spanish!

Special note: Our Company First Sergeant, S-4 Section Sergeant, and our Ammo Supply NCO, we're all told that "They Were Not Needed" and would be remaining behind. Other unit members were also told this.

The S-4 RTO (Radio Transmission Operator) is a support position, not a combat position like say a Radio Transmission Operator of a rifle platoon. The S-4 Captain had no mission; he was just there to give support when needed, such as re-supply. And I was to follow him around, in case he had need of a radio!

"I was never given a mission briefing," only told that we were jumping onto the international airport at night, and that I was to link up with the S-4 Captain on the ground.

We all had to stand in line for shots, and then returned to our temporary unit area to pack!

While preparing for our combat night jump into the country of Panama, we were all packing our rucksacks (backpacks) with the equipment assigned to us for our war time duties. When the S-4 Lieut. walked up to me with two armloads of equipment and said, "Cloak, put this in your rucksack. I don't have room for it!"

The RTO equipment I already had in my rucksack was very large as it was the old Vietnam era radio, along with two extra batteries and accessories such as the antenna and microphone.

My rucksack was totally full. So I said, "Lieut. my rucksack is full. Take this equipment and give it to someone else who does not have the same load I have. I am the acting S-4 RTO! I have to be able to keep up with the S-4 so that his radio is within his reach at any given moment."

The Lieut. snapped at me saying, "I don't have room in my rucksack! You put it in your rucksack and carry it for me! That's an order, Cloak!"

I then had to completely empty out my rucksack and leave my own personal items such as extra socks, extra uniform, extra water and completely take apart my three days worth of MRE's! Then stuff the individual packages into the outside pockets of my rucksack/backpack, crushing them into place. Only then could I get the lieutenants equipment packed alongside the radio inside my rucksack/backpack.

We then flew for many hours in a C-141 Star Lifter for our nighttime mass tactical assault into the country of Panama. As I exited the door at 500 feet over the international airport in Panama City Panama, red tracer rounds from AK-47s were buzzing by me and going up through my parachute.

Normally, there are 4 regular rounds before each tracer round. That means for every 1 red tracers round you see, there has been 4 regular rounds before it. I seen about 20 go past me up into my parachute! Which adds up 5x20= 100.

With the deafening sound of jet engines, I looked below me to see the outline of a C-141 Star Lifter passing by in the darkness. This was a horrifying event; I thought I would be hit by the next C-141 that was still coming! Until I realized "by the starlight", I was looking at the shadow of the C-141 Star Lifter that was going by, over the top of me, and I was about to hit the runway.

Grabbing the release strap that held my rucksack, I gave it a strong jerk, releasing my rucksack and weapons case to fall the length of the 30ft cord they were attached to.

Before it reached the end of the 30ft cord, I heard my rucksack hit hard on the concrete. I then hit the runway and twisted my right angle as bullets hit the concrete runway inches from my head.

Pulling on the cord, I found my equipment not far from me. Opening my weapons' case I pulled out my M-16A2 and took aim over my rucksack. In between me and the flash of enemy weapons were friendly soldiers. Then suddenly another C-141 Star Lifter passed over head, dropping more airborne infantry soldiers. They were landing everywhere!

I could not return fire, because members of my unit were in between me and the enemy. It was after midnight and O Dark 30. I would not take the chance of killing my own unit members and because bullets were still hitting all around me, I low crawled into the high weeds behind me and out of sight of the enemy soldiers who could see me as bullets were still hitting all around me.

[Not firing through a friendly soldier was deeply engrained in my consciousness, from the long range patrolling course I had gone through while I was in the Special Forces Qualification Course. Because, while in the long range patrolling course and in a listening holt, we were attacked from my rear. Which was on the other side of the complete 360 circle, we were all lying in, and I had turned around firing M16 blanks across and through my unit toward the attackers? Thus killing my own people and destroying any favorable outcome to the mission. (I was yelled at by the "Green Beret" instructors until I knew better than to ever do this again, no matter what!]

I heard bullets hit in the weeds near me as the friendly soldiers nearest the terminal building returned fire and 1 then 2 enemy positions went silent.

When the bullets stopped hitting near me, I low crawled, still in the darkness, until I found other friendly soldiers not in my unit. Soon, more came and we began to form up together in the darkness. Then once the sun came up, I found my unit and rejoined them. The radio never worked, the hard hit on the concrete runway had taken its toll.

If you are connected to the internet and have Google Earth setup on your computer, you can find the exact spot, I landed on the runway and the exact spot, I low crawled into high weeds.

Go to Google Earth: Panama City Panama, I landed at 9 degrees 03'50.53" N by 79 degrees 22'59.44" W and 9 degrees 03'48.36" N by 79 degrees 22'56.40" W was where I low crawled to.

As the acting S-4 RTO, I was in a support position and because the radio was not working, I was of no use to the S-4 Captain. He turned me lose, so that I could commandeer a truck to transport water up and down the runway of the International airport to our battalions different units who were positioned in the hot sun and running out of water. They were gratefully happy for this needed water and soon moved out in different directions, going on their combat missions!

Three days later, we were resting after a mission we had been on, where we confiscated a large amount of weapons from an arms' room. These weapons were captured at the Panama Defense Force, Battalion 2000 unit area.

I had taken a bulletin board full of Panama Defense Force, Battalion 2000 member's photos from one of their offices back to the International Airport. Where I found prisoners who had not been identified as members of the Panama Defense Force, Battalion 2000, and pointed them out to the MP's; enabling them to separate out the officers and non combatants!

I had also taken a brand new "AK-47 Bayonet" while helping to empty the arms' room, it was still in its packing from the factory in Russia, and put it into my butt pack on my equipment belt. This would be my war time treasure that I would show to my grand children in the future.

I also commandeered an old brown flak vest, from the Panama Defense Force, Battalion 2000 unit area and was now wearing it.

After turning over the bulletin board full of Panama Defense Force, Battalion 2000 member's photos to the MP's, I noticed people waiting by a phone, so I stood in line!

There I called "My Bubby's Grandmother" in Panama and asked if everyone was ok! Then asking her to please make sure that all of the children remained inside the house, as I knew they would want to run and see whatever was happening and could get shot or killed.

I was worried about their safety as the fighting had been going on for 3 days, and I had not been in touch with them for 3 months!

I then went to lie down next to my rucksack as I had not slept for many days. When I was asked by the S-4 Lieut. "Did I know how to get to the Marriott Hotel?"

I answered "Yes, I have been there many times!"

He said to follow him, so I did! A minute later, I was standing among the S-4 Captain, the Delta Company Commander, The Delta Company First Sergeant and the S-4 Lieut. The Delta Company Commander was complaining about his being given a mission to get ammo, water and food to one of our battalions platoons that had been cut off from re-supply for 3 days. Saying that he did not know how to get there!

I spoke up and said that "I had just spent 2 ½ years in Panama and knew how to get there easily." The S-4 captain then asked for me to show the Delta Company Commander; "How to get to the Marriott Hotel."

I pointed to the map as I explained that all he had to do was, get onto the highway next to the international airport and follow it west until he came to the Roosevelt Monument Intersection. Then to turn left and follow the same road until he came to the ocean and the "Spanish Fort Ruins," then continue to follow the same highway as it took a right turn along the shoreline. Continue a few miles along side of the shore line, and then he would see the Marriott Hotel.

Go to Google Earth: Panama City Panama international airport, 9 degrees 03'50.53" N by 79 degrees 22'59.44" W.

The Highway: 9 degrees 04'13.64" N by 79 degrees 23'21.42" W.

The Roosevelt Monument Intersection: 9 degrees 01'52.96" N by 79 degrees 29'25.28" W.

Turn right at the ocean, "Spanish Fort Ruins": 9 degrees 00'23.96" N by 79 degrees 29'10.51" W.

Marriott Hotel at that time in history: 9 degrees 08'59.20" N by 79 degrees 30'06.17" W.

Then I pointed on the map at the "Spanish Fort Ruins" and said, "The last time I was there the (PDF) Panama Defense Force had their people stationed inside this building on the corner, and they were patrolling the area with AK-47s. You should get on the radio and get air cover so that when we get there, we are not ambushed. They can easily see us coming from a mile away!

The S-4, Lieut., spoke up and said "No, that's ours! Once we get there we're safe!" I said again, "The last time I was there, the PDF had it, and they had AK-47s in clear view." Again, the S-4 Lieutenant spoke up and said "No, I heard on the radio that's ours. Once we get there were safe!"

Again, I repeated myself and said, "Don't you want to, just get on the radio and make sure? Look here on the map, they can see us coming from a mile away and set up a hasty ambush."

Again the Lieut. said "No! That's ours! I heard it on the radio. Once we get there we're safe. Can you get us to there?" I spoke up and said "Yes, I can get you there easily. Even so, don't you want to get on the radio and call to make sure?"

Again the Lieut. said "No, I know it's ours!"

At that time, the Delta Company Cmdr. said, "I don't like it! If I make a wrong turn, and we get down a dead end street we can get ambushed and killed. I'm not going to do it!"

I then spoke up and said, "I can go along with you and guide you. Look here on the map," as I showed him once again. The Delta Company Cmdr., spoke up and said "You would do that?" I said again, "Yes, I volunteer to be your guide!"

The S-4 Captain spoke up and said "Well there you go, you have a guide!" The Delta Company Cmdr. said "Okay, if you'll go along with us as guide, I'll do it!"

I said, "I will." The Delta Company Cmdr., then said, "Okay, let's go!"

The Delta Company Cmdr., then called over to the Sheraton tank Cmdr. and told him, "You take the lead; we will pull in behind you." The Sheraton tank Cmdr., spoke up and asked, "What do I do, if I come upon a roadblock?"

The Delta Company Cmdr., then replied, "Don't stop for anything, just bust it down and keep moving. Don't stop for anything!"

I went to get my rucksack. Then walked over to the lead Humvee in the convoy and started to sit down in the front seat, where I needed to be as the guide. At that time, a platoon sergeant spoke up and said "No, you sit on the back. I spoke up and said "No, I need to sit in the front, I'm the guide! He spoke up again, saying "No, get in the back!" I repeated, "I need to be in the front, so I can give directions to the driver." He again said "No, you ride on the back!" I then walked to the back with my rucksack putting it over inside the back of the Humvee, next to the stack of MRE's.

As I walked to the back of the Humvee, a soldier, maybe the driver said "I have a M203 Grenade Launcher, and you have an M16!

The fighting position you will be in, in the back, is a M203 position. Where I am, is a M16 position! If we get into a fight, you will need a M203, and I will need an M16! So let's trade off, until the mission is finished! I said "YES!" and we traded off knowing that we would find each other after the mission and trade back our weapons!

I then started helping soldier's pick up the pile of rucksacks that were on the concrete runway at the back of the Humvee. The S-4 Lieut. walked up and put his rucksack down in the pile to be loaded on the Humvee.

Each rucksack I picked up, felt like it weighed between 80 and 90 pounds. So I had to grip it firmly and give a good solid jerk as I picked it up and handed it to the soldier in the back of the Humvee.

After picking up five or six of these heavy rucksacks, I came to the rucksack that the Lieut. had laid down. Expecting it to be as heavy as all the other rucksacks, I grabbed a hold of it firmly and gave a quick jerk, so that I could lift it up to the soldier in the back of the Humvee.

It was so light that I easily lifted it over my head! Shocked at how light it was, I bounced it up and down in my hands. Much like you would do a basketball or a football. I looked at the soldier in the back of the Humvee and said "This rucksack is as light as a feather. It looks like it's full, but is packed full with a pillow!"

At that point, the S-4 Lieut. walked up. So I turned to him, and while bouncing the rucksack in my hands I said, "This rucksack is yours, right Lieut.!"

He became very quiet and wide eyed! I then said, "This rucksack is as light as a feather, what have you got in here, a pillow?"

He spoke up with a derogatory commanding voice and said, "What do you think, I'm stupid. I have to run!"

I was instantly insulted and started to argue with the Lieut., to find out who else he gave out his equipment to, so he could pack his rucksack full with a pillow, insuring that it looked full, to anyone that seen it! Still light enough he could run unburdened if he ran into an enemy!!

I heard someone yell, "Okay, let's go people, let's go!" Everyone started jumping into the back of the Humvee. The Lieut. took off running in the direction of another Humvee.

As I climbed up in the back of the Humvee there was only one small place for me to sit down, which was on top of a stack of MRE cases, on the right side of the rear of the Humvee. This left me totally exposed, as I was sitting up higher than anyone else and in the back of a turtle back Humvee. There was just barely enough room for my knees to fit in under the roof behind the person in the back seat on the right side.

The trip was basically uneventful, as I guided by yelling my directions down to the person in the front seat, which was a platoon sergeant, and then he relayed over the radio my directions to the Sheraton tank Cmdr... They had taken the lead; we were now the second vehicle.

At one point, I spotted a small tank on a hill to the right and called it down to the people in the Humvee. I heard it repeated over the radio microphone! Tank On The Hill, To The right!

As we came up on the intersection of the Roosevelt Monument, I called down my directions and said "That's the Roosevelt Monument intersection, turn left!" and the Platoon Sgt. repeated it over the microphone; "That's the Roosevelt Monument intersection, turn left!"

We then turned left and begin heading in a southern direction!"

As we came to the first intersection after turning left at the Roosevelt Monument, we came upon a roadblock that consisted of 55 gallon drums cut in half and filled with diesel fuel. Then caught on fire!

As we neared them, the Sheraton take skidded to a stop, which caused us to come squealing to a stop behind them. Then, each individual vehicle behind us had to stop, which is called the slinky effect!

As the platoon sergeant urgently screamed over the microphone "Don't stop, keep moving! Don't slow down, keep moving! The Sheraton tank took off, and so did we, bouncing over the top of the squished drums and diesel fuel, inflamed, now all over the street.

Then we came up on a side street to the left and again a roadblock that consisted of 55 gallon drums cut in half filled with diesel fuel caught on fire.

Again, the Sheraton tank came skidding to a stop. At the same time, the platoon sergeant yelled into the microphone "Don't stop! Don't stop! Just run over it! Don't stop for anything, keep going!"

We narrowly missed hitting the back of the Sheraton tank as we swerved to the left and over the top of a burning drum of diesel fuel, which got hooked up under the Humvee and splattered fuel all over the tire that was directly under me.

As the Sheraton tank picked up speed it got back in front of us. We pulled in behind it, and soon came up on an intersection and a roadblock that consisted of 55 gallon drums cut in half and filled with diesel fuel. Then caught on fire!

Again the tank stopped and we had to stop behind it!

When the tank moved on forward over the drums, we also ran over it! The drum that was caught under our Humvee from the previous roadblock had broken free, falling away. We moved on with the right rear tire under me still smoking from the fire.

Then we came to a road entering from the right of which a small car pulled up to, stopped and two men started getting out of the car with AK-47s.

We continued on passed them as I tried to turn to the right and point my weapon in their direction! It was too late; they were now out of my view!

Immediately, something hit me that felt like it was a 50 pound rock the size of a grapefruit. Just below my left hip on the outside!

I thought maybe somebody had thrown a big heavy rock as I had heard no sound over the roar of the tank engine and the Humvee motor. The men with the AK-47s were now behind me. So I looked down, in between my legs on the floor where it would have fallen.

I could see nothing!

As we continued to move forward, I looked to the left, because my lower hip was hurting like hell on the left side. Which is where I thought whatever hit me, must have come from?

There, I seen people in the side street waving white handkerchiefs and yelling the word "Liberators." At that instant, I believed what hit my leg "may have been a grenade" and again bent my head and eyes forward to look down between my legs on the floor of the Humvee to find it.

Right then, at that exact moment, everything went black!

A black tunnel had quickly closed in on me, I was passed out cold!

The spot where the bullet went into the upper inside of my left leg and exited near the middle of the outside of my left hip, scientifically proves that the shooter was on or near a 3rd floor of a building. And was most likely aiming at the Platoon Sergeant sitting in the front of the Humvee ahead of me (We were moving!), as his head was at the same level of my gunshot wound.

Remember that I was sitting in the back, up on top of cases of MRE's looking over the top of the Humvee and that puts my hips at the same level as his head and directly behind him, by about 3 1/2 feet.

To find the place I was hit and then passed out, go to Google Earth, Panama City Panama, point your mouse at 9 degrees 01'29.45" N by 79 degrees 29'15.39" W and then zoom in.

I passed out at Google Earth: 9 degrees 01'23.77" N by 79 degrees 29'14.32" W

I awakened as if I was in a tunnel and off in the distance, every weapon in the world was firing bullets. It sounded like I was on a firing range back at Fort Bragg.

Then I heard someone calling my name saying "Sgt. Cloak, which way? Sgt. Cloak! Sgt. Cloak, which way?"

I woke up at the "Spanish Fort Ruins" / Google Earth: 9 degrees 00'23.58" N by 79 degrees 29'10.53" W, so you can see that I was passed out for 2 to 3 minutes as that is how long it would take to drive that far!

I have no memory of it!

What fighting went on during that time period, I have no knowledge of!

Slowly, my eyes began to focus as the sounds of bullets were hitting the Humvee, seemingly from every direction all around me.

I turned my head and eyes to the right as an American soldier came running up to the wall on the side of the street of the old Spanish ruins! As I continued turning my head and eyes to the right, I could not twist through the pain in my hip far enough to look behind me down the street!

Right in front of me, the man standing in the hole in the top of the Humvee for a Gunner was firing his weapon off towards the PDF building, to our forward left. It was the building I had pointed to and described on the map that was the "Panama Defense Force," guard shack or guardhouse at the "Spanish Fort Ruins." The one I had asked, "Why don't you just call on the radio and confirmed that it's ours! Because they will see us coming from a mile away and ambush us."

The man to my left on the back of the Humvee was pointing his weapon to the left front, and firing!

To the forward left was the only opening to fire through!

At that moment, I pointed the M-16 with 203 grenade launcher and started firing rounds into the window on the side of the building, the weapon jammed.

After changing magazines, and un-jamming the weapon, I fired many rounds into that window. Then I took a high explosive round and put it into the M203 grenade launcher and fired it at the window hitting the side of the building all while sounds of everyone's weapons were firing all around me.

Again the voice yelled, "Sgt. Cloak, which way? Sgt. Cloak, which way?"

Looking to our front, in front of the Sheraton tank, I could see that there was a "School Bus" on its side, lying diagonally across the highway blocking our convoy.

I yelled, "Around the bus, turn right!"

I was later told that the tires of that Hummer had 103 bullet holes.

Immediately, the "Platoon Sergeant" repeated my command, "Around the bus, turn right!"

The Sheraton Tank began pushing the school bus off the road as if it were a bulldozer bent on destroying anything in its path.

Once the "School Bus" was out of the way, our Humvee took off like the speed demon it was meant to be.

America, you can hold your head up, and you can hold it high! That Humvee engine kicked into overdrive and came around that curve like a 4 wheel drive dragster out of hell.

It truly was a scene, any patriotic movie director would be happy to film, with an "American Bald Eagle, Soaring Over Head" while screeching a warrior's cry!

PTSD Bald Eagle,
Screech it loud and clear!
"Let go, your warriors cry, Bald Eagle, Let go, your warriors cry!
Screech it loud and clear, Bald Eagle, Screech it loud and clear!
American blood flows bright it's clear, American blood has been shed right here!
Let go, your warriors cry, Bald Eagle, Let go, you warriors cry!
Screech it loud and clear, Bald Eagle, American blood has been shed right here!
KKeeeeeeeeeeeeeeeeeeeeeeeeeeeeeeerrrrrrrrrrrr"

As tears fill my eyes, I remember this heroic event, never before told to the American public!

The soldiers still able to fight were returning fire like the professionals they were trained to be!

American soldiers were bogged down in an ambush, and we needed to get out of their way!

They needed to move out of the kill zone!

We did, and that Humvee engine continued to roar its way down the street. Suddenly, out of nowhere, another Humvee came alongside of us and had to slow down to keep from passing us.

How, I don't know!

We were flying!"

Someone was hanging out the side with their head split open from a gunshot wound. The person next to him in the Humvee was holding on to him so that he did not fall out onto the street.

He was out cold!

They were yelling "We have to stop! We have wounded, we have to stop! We have wounded, we have to stop RIGHT NOW!"

Immediately tires were screeching to a stop, while soldiers were jumping from the Humvee's in all directions. As I twisted so that I could turn and climb out of the Humvee, a pain shot from my left upper thigh, lower left hip, up into my brain that said, "DON'T YOU EVER TRY TO DO THAT AGAIN! NO NOT EVER!"

As I looked down toward my left hip where the pain was coming from, I could see nothing! It was directly under my ammo pouch on my cargo belt. As I reached to pull it out of the way, what I saw was a spot in my uniform trousers that looked like it was ripped, but also red with blood!

100% American Blood; mixed through generations going back before the civil war between the states!

It was exactly where my leg creases, when I am sitting down! The pain was so intense that I sat looking through black spots, trying to focus on the highway ahead of me.

Everyone had disappeared, running off away from the Humvee in a direction off to my right rear, out of my sight!

I was alone and could not twist to turn and see where they had gone! Again the sounds of weapons firing took over my mind as I looked through the black spots, which were now growing larger!

I felt as though I was going to pass out, as the black spots in front of me were becoming larger and larger by the instant; it was my energy draining away!

Then, someone walked over to me beside the Humvee and said, "Come on, get down!"

Trying to focus on his face, I replied; "I can't, I'm shot!"

He climbed up on the side of the Humvee and said in a disbelieving voice, "Where are you shot!"

The disbelief in his voice gave me a little energy, so I pulled back my AMMO pouch on my left side, and said while pointing, "Right there!"

He leaned forward and looked; His eyes opened very wide as he focused on the exit wound and blood. Then he said loudly, "Oh My God! Don't move; I'll be right back!"

Seeming like an eternity later, 3 men gathered around me and twisted me out of my sitting position on top of the cases of MRE's, ripping the torn flesh where the bullet had gone through even further!

They had pulled my legs out from between the top of the Humvee and the case's of MRE's, handing me to someone on the ground.

The pain from twisting my leg while pulling me out was causing me to black out again! They then put me into the back seat of the Humvee on the right side and strapped me in with the seat belt.

Sounds of weapons firing sent everyone running, again to my right rear and out of sight.

As I fought through the black spot's engulfing my vision, the sounds of weapons firing in the near distant behind me out of my sight, began to slow and then stop.

Someone was yelling at me to get out of the Humvee! Each time I pushed on the release button on the seat belt buckle nothing happened, it would not release!

I could not focus my eyes through the black spots before me, but could feel the seat belt buckle on my right side. My hands felt the button as I pushed it, it just would not release!

Confused; I watched as our battalion surgeon, a Captain; walked up to me, while demanding, that I get out so that he could take a look at my gunshot wound!

I did not know he was with us! I had not seen him sense we left our battalion area back at Fort Bragg, days before.

Now, seemingly out of nowhere, he had just walked up to me out of my peripheral vision, angry that I would not follow orders, to get out of the Humvee. So that he could examine me.

Again, he yelled; "Get out here and lay down, so I can take a look at you!"

As I tried again to push the release button on the seatbelt buckle, I said "I cannot get the seat belt off!"

He then tried and when he could not get it to release, he took out his surgical scissors and cut the seatbelt in half! He then helped me to get out of the Humvee, and to lie down on the ground!

At once, the world up roared around me as weapons began firing seemingly from every direction.

I could not roll over, so I tried to look back in the direction of the bus that had been blocking the road minutes before by tilting my head back as far as I could see.

At that moment, someone had a hold of my collar and was pulling me along the ground with them. The whole time, weapons were firing all around us; the fire fight was not over.

I kicked with my right leg and foot, trying to help as the faceless soldier pulled me along the ground. Each time I tried to use my left leg, the pain called out to me throughout my brain saying; "I TOLD YOU NEVER TO DO THAT AGAIN! NOW LOOK WHAT YOU HAVE GONE AND DONE!"

Soon I would be pulled and dragged over behind the walls of the Old Spanish Ruins, as our battalion surgeon was trying to save the life of a soldier that was shot much more seriously then I had been! He had been in, even another Humvee!

As soldiers around me cut and pulled at my uniform to get it off of me, allowing the medic's to see all wounds. I struggled against them not understanding why they were undressing me. While weapons began firing, once again, seemingly from everywhere around me!

Unable to keep them from removing the flak vest, I turned my attention to my "Bullet Proof Kevlar Helmet" by rapping my arms around it and not letting go.

I was out of my mind with confusion, while fighting black spots; as to why they were totally undressing me in the middle of a fire fight! I felt the need to point my weapon and pull the trigger, but had good guys holding me down while cutting and pulling at my clothes!

It didn't make any sense!

Striating out my leg had further ripped the muscle tissue in my lower left hip where the bullet had torn through it. Causing the pain to over whelm my mind, leaving me shaking and cursing until someone gave me a stick to bight on!

I bit down hard on the stick as the pain worked to over whelm my thought process and focused on the face of a "Red Headed Sergeant", I remembered from my days in the 505th, noticing a kind understanding look in his eyes, while wondering how he got there!

Suddenly, he pulls out from with-in his shirt pocket, a small sucker or lollypop, put it in front of my eyes and says; "I'll trade you this for that stick!"

The calm understanding in his voice, reassured me! So I took it, while taking the stick out of my mouth and said, "Thank you! I don't understand why they are taking my clothes off!" He only smiled!

Soon the candy was gone, and I was back to biting on the stick!

My concern was not motivated by fear, but by pain and confusion. Including my desire to respond in the correct manner! I was thinking about every little thing that was happening, considering each alternative, ultimately the answers never came to mind. It was just too painful, too confusing!

Over to my left, I could see our battalion surgeon working on a soldier who was unmoving! The world had gone silent!

Soon, I and 2 other soldiers, 1 shot in the arm and another unmoving, were all medevac-ed out in a Black Hawk Helicopter. We landed at a M*A*S*H unit on the runway at Howard Air Force base next to Fort Kobbe Panama.

On the way there, the soldier who had been shot in the arm said; "I got him Serge, he was in a window! I opened up on him, he went down!"

The rescue mission would continue without me as guide!

I was left in the helicopter once it landed, so that emergency procedures could be performed on the more desperately injured soldier.

He passed away!
May God Please,
Bless his Mother and Father!
Amen!

PTSD Bald Eagle,
Screech it loud and clear!
"Let go, your warriors cry, Bald Eagle, Let go, your warriors cry!
Screech it loud and clear, Bald Eagle, Screech it loud and clear!
American blood flows bright it's clear, American blood has been shed right here!
Let go, your warriors cry, Bald Eagle, Let go, you warriors cry!
Screech it loud and clear, Bald Eagle, American blood has been shed right here!
KKeeeeeeeeeeeeeeeeeeeeeeerrrrrrrrrrrr"

Three hours after being shot, it was dark now; a member of the M*A*S*H unit dressed in all white, walked over to me with a very large clear syringe. He took the stick out of my mouth throwing it away; I was still shaking!

The clear syringe was the size of a Coke can, filled with a clear solution. It had a large plastic point on the end of it with a large opening. The M*A*S*H unit member pushed it into my exit wound, injecting the entire contents.

I screamed bloody murder, for the burning pain was more intense, then the moment I was shot.

The man in the white shirt grabbed a hold of me, as if he thought I was going to rise up to start running.

Then he said "Oh my God, Oh my God! I thought you already had some morphine! That was pure alcohol!"

Angrily, I spoke up and said "NO, I HAVE HAD NOTHING!"

He then said "Wait right there, I will get you some morphine!" Once I was given a shot of morphine, my lights went out. I woke up feeling like I was inside of a freezer.

As I looked around me, I could tell that I was inside of an army bus with a blanket over me. However, my bare feet were sticking out into the cold air! I would soon find out that I was in a parking lot at an Army Air Base outside of Ft. Sam Houston, Texas.

The next day, I was given more morphine and a spinal anesthesia so that the surgeons could cut out the dead or dying flesh of my gun shot exit wound.

On the Internet Wikipedia states: Spinal analgesia, (or commonly called spinal anesthesia or sub-arachnoid block (S.A.B.)) is a form of regional anaesthesia involving injection of a local anaesthetic into the cerebrospinal fluid (CSF), generally through a fine needle, usually 3.5 inches long.

A round hole large enough for a grown man to put his fist into! Was left to drain for days, before it had to be elongated by cutting 2 large V-shape cuts, at the left most and right most portions of my gun shot exit wound! While sitting down, that is the upper most and lower most!

This had to be done, because the perfectly round hole, cut by the surgeons could not be closed. The 2 large V-shape cuts which elongated the exit wound, allowed the surgeons to sew shut my gaping war wound after another spinal anesthesia.

This left me with server "spinal headaches" caused by the "Lumbar Puncture" into my spine," and me to hobble around on crutches for months to come.

A few days later, I and all other wounded soldiers in our hospital ward were visited by President Bush, the 41st President of the United States.

The day before President Bush and the First Lady showed up at the hospital to visit everyone that had been wounded on the same floor I was on. Men started showing up that were obviously Secret Service!

This one particular man! Who stood by a window near me? Watched out into the distance, including, down to the grassy area besides the building... He was wearing a suit that you might think was that of a family member, not a Secret Service member.

He was obviously Secret Service as his activity told me exactly what his job consisted of! Plus, he didn't talk to anybody that was in a bed, who had been wounded in combat. That's only confirmed my suspicion because of the jacket of his suit; being more like a common man's jacket, then that of a Secret Service man's jacket.

Because of the "spinal headaches" causing me to pass out each time I lifted my head off of the pillow, I was not able to stand when "President Bush" shook my hand or walked around the room with the "First Lady," meeting other more tragically wounded soldiers like my friend Sgt Alameda! Who had, had 109 pieces of "Grenade Fragments and Bone" removed from his body because of a grenade thrown by an enemy soldier onto him and the two soldiers with him who lost their lives.

May God Please,
Bless their Mother's and Father's!
Amen!

PTSD Bald Eagle,
Screech it loud and clear!
"Let go, your warriors cry, Bald Eagle, Let go, you warriors cry!
Screech it loud and clear, Bald Eagle, Screech it loud and clear!
American blood flows bright it's clear, American blood has been shed right here!
Let go, your warriors cry, Bald Eagle, Let go, you warriors cry!
Screech it loud and clear, Bald Eagle, American blood has been shed right here!
KKeeeeeeeeeeeeeeeeeeeeeeerrrrrrrrrrrr"

My daughter was there with me; my parents had brought her to see me! My life had taken my wife and "My Little Bubby from a 3rd world country, all the way up to meeting the President of the United States; as they were there with me when the President arrived!

I felt bad, that there was not enough room for my father and mother to stay! We were told that my family was too big, someone would have to go!

After photos with "President Bush" and that precious moment of shaking his hand and he had left! I needed to go to the restroom. So badly that I forced myself to face the horrific headache and nausea as I hobbled my way to the restroom! Just then, the Secret Service man was in the restroom washing his hands so I spoke up and asked him, how I could get a job with the Secret Service?

Looking me straight in the eyes he replied, "You don't want this job!" I replied, "Yes, I do!" He said again more sternly, "You don't want this job!" Again I said "Yes, I do, I would love it!"

Very seriously, he says, "No! You wouldn't like it! I'll tell you what I'll do! I'll give you a number to the Pentagon! You call this number and tell them I gave you the number and said for you to call them!"

I thanked him and took the piece of paper with the phone number back to my bunk and nearly passed out from the intensity of the headache I was in during.

Did he know something, I didn't know? Apparently so!

I was then taken into an operating room, again given a high dose of morphine. A spinal blood patch had to be done. To close the hole in my spinal fluid sack! Which was left open after I was given the "spinal anesthesia" before my operation when the doctor had elongated my gun shot exit wound? Now, the cause of my spinal headaches!

They did the blood patch by taking a large amount of blood from my arm, injecting it into my spinal fluid. Around the area of my spinal anesthesia injections! This caused my back, to lock or harden like a stiff rod of iron, hurting like hell for many days!

However, it did stop the "spinal headaches," that were caused because each time I tried to lift my head or stand up, gravity pulled on the spinal fluid in my spine; causing it to leak out though the holes left in my spine from the needles; causing a type of suction on my brain!

I was told that the blood would clot as it ran out the hole in my spinal sack and help it to heal faster.

Soon I was awakened, asked to answer some questions for a magazine reporter. I did, as I was feeling very high from the morphine.

It makes life a party!

A few days later I was flown to "Womack Army Hospital" at "Fort Bragg," being put into a ward there! Two weeks later, I was given crutches, and then sent back to my unit area. Once there, I was handed a set of cargo belt and shoulder harnesses with a butt pack. Told that it was mine, sent from Panama.

When I opened it, the "AK 47 Bayonet", I had put into it was not there. It was not my butt pack. I was highly disappointed not to have my war souvenir! Even so, a 45 caliber magazine loaded with military rounds was there. So, I was happy to keep it as my war souvenir!

(Although years later, when I had taken my "Bubby" to the country of Panama. To meet his grandmother for the first time! My niece's husband got into my things, and sold it for the money to buy a "6 pack of beer!")

When I had arrived back at my unit area for the first time, the driver's side window was smashed out of my car. I had left it in the parking lot, and had given the keys to the Ammo NCO of the S-4 Section, who had been left behind and told "He was not needed during Operation Just Cause," so that he could give them to my wife when she arrived, to pick up the car.

He never had given her the keys and was walking around with his right fist in a cast up to his elbow!

Did your eye's just open wide?

They should have!

Two plus Two, equals Four!

While there I was approached by the First Sgt., who had been left behind and told "He was not needed during "Operation Just Cause." He complained about my being away for my job as "Air Items NCO" and wanted to know when I would be back.

I explained that I was still on bed rest, but had been sent home, and that I was at the company area, because I wanted to talk to the S-4 Capt... The S-4 Capt. was nowhere to be found, he was still in the country of Panama!

Two weeks later, I went to my chain of command to talk about what the S-4 Lieut. had done by passing out his equipment to me and the other soldiers so that he could fill up his rucksack with a pillow.

This took me to the S-4 Section Sergeant, an E-6 that was told he was not needed during "Operation Just Cause" and was left behind. As I explained to him what took place, he sent for the Lieutenant… Once the Lieutenant arrived, I expected, the S-4 Section Sergeant to have the same feelings I had about the situation.

However, he did not, in fact, he teamed up with the Lieutenant and the First Sgt., relieving me of my duties and assigning me directly under the First Sgt., inside his office as a clerk.

I was soon standing in front of the S-4 Capt. being questioned about my interview with the magazine reporter. He was very angry and upset about what the reporter had put in a magazine article, which was my exaggerated value of the weapons, we confiscated out of the PDF Battalion 2000 arm's room, and my phone call to "My Bubby's Grandmother!"

I explained that I didn't remember much about my interview with the magazine reporter, and that I was high on morphine during that period in the hospital, and I answered yes, that I had told the S-4 Lieut., about calling my mother-in-law!

The next day, many soldiers that walked by me inside the company area, were making comments like, "I heard you got shot by friendly fire!" It was always someone that was no wear near me when I was shot or was left back at Fort Bragg during "Operation Just Cause," told that they were not needed!

From that time forward, I was treated rudely by all rank levels. The First Sgt. even went as far as ordering me to falsify his PT records by changing his push up count from 38 to 90 and his sit-up count from 32 to 89. Then he watched to make sure, I did it correctly.

The duties I was responsible for as a clerk in the First Sgt.'s office, was to keep all jump records updated for the 100+ men in our airborne infantry unit. Many times, the First Sgt. sent me on errands. One particular time, he sent me to the PX to buy him a box of "Chocolate Gummy Bears!" Then when I returned, my little "Bubby's" mother, my wife, was sitting in his office talking to him.

When I entered and asked her how long she had been waiting for me. She stated that she had just arrived and needed money to go shopping with. I asked her were my little Bubby was, she said that she had left him with a neighbor.

All through The History of the 82nd Airborne, a clerk in the First Sgt.'s office is exempt from pulling 24-hour duty. The First Sgt. informed me, that it no longer applied and had me pull 24-hour duty at least once a week.

I had earned a Purple Heart, 2 Bronze Stars and a Combat Infantryman's Badge, along with an Expeditionary Medal with Spearhead, for my actions during "Operation Just Cause." But was never given an award in formation or allowed to participate in any award ceremonies.

My "Purple Heart" Metal was simply left on a pillow while I was sleeping in the hospital after being wounded. My 2 Bronze Stars, Combat Infantryman's Badge along with an Expeditionary Medal with Spearhead was simply handed to me by use of a piece of paper. Which were the United States Army orders, awarding them to me?

By July 1990, I was no longer using my crutches and had begun to exercise with the headquarters platoon at my own pace. Late one afternoon after returning home and taking off my boots and socks! I walked over to sit down at my lazy boy chair to watch TV. When I stepped on something in the carpet, which sent a stabbing pain throughout my entire body! Something had driven its way all the way to the bone on the pad of my left foot just under my left big toe.

Sitting down in my chair I lifted my foot up, to see a thick posted earring. The type that women use to wear back in the 1960s! My wife immediately started laughing! I did not think it was funny as it hurt intensely.

I asked her, whose earring it was! She claimed it was hers, while I claimed that I didn't know she had any earrings. She never wore one before! I didn't feel good about the way she was laughing about it, but blew it off, as it must've been funny from her point of view.

The next morning the spot where the earring went into the pad of my foot was burning. So I put a large Band-Aid and some salve on it. Once at my company area, I went to the medics, they gave me some Tylenol.

Three days later, it was so bad, that a red bubble the size of a quarter had been swollen around the puncture wound from my stepping on the earring. I went to the medics again; they sent me to the hospital, saying that I had blood poisoning.

At the hospital, I was admitted and put into the Ward at the end of the hall. Where I walked to years before when I had lockjaw from my dental visit, and wanted to find the obese nurse that woke me by using my index finger to masturbate herself.

After 5 days, the blood poisoning had started to move up around the side of my foot. So I called home to speak to my wife and ask why she had not been to the hospital to visit. She claimed she could not find anyone to drive her there or to watch my little Bubby!

The Dr. had taken samples from the puncture wound and sent them off to the laboratory, mean while he was giving me penicillin.

When I had been in the hospital for 9 days, and the blood poisoning was continuing to crawl up my foot toward my ankle. The doctor changed to different penicillin and a much higher dose. Exclaiming, that the laboratory had found no reason for my blood poisoning! There was simply nothing there to cause it.

Five days later the blood poisoning had almost reached my ankle on the top of my foot. The doctor told me that he was going to mix up a cocktail, of every known form of penicillin! Then add it to my IV for the next 24 to 48 hours!

He said, "If the blood poisoning has not stopped moving up your foot within two days, I will have to amputate your foot!"

This caused me to call home to my wife, and ask that she come by to visit me with "My Little Bubby." Also to call my family and let my parents know what was going on. She still could not find anyone to drive her there as she did not know how to drive!

It was past time to pray to my God, so I did, and I did it by the name of "Jesus Christ of Nazareth" asking that the penicillin cocktail would work!

During my examination two days later the doctor said; that the penicillin cocktail was working and the blood poisoning had stopped moving up my foot toward my ankle.

I was in the hospital for another week before I was allowed to go home on crutches with a week on medical leave!

I then re-enlisted for 2 more years as my current enlistment was up, and I was in no shape for ranger school at that time.

A week later I was in the country of Saudi Arabia performing duties during the first few days of "Operation Desert Shield!"

The day after I had reported seeing the man that looked like Osama bin Laden, started the 2 months that the Saudi Arabian government would only give us 1 hot dog bun, 3 times a day to eat, because we were barbarians. It caused me to lose a lot of body weight, which I would not regain until after leaving the US Army a year later!

2 weeks later, while running in formation with my unit. A Lieut. ran up beside me and said "Cloak, why don't you just put a gun to your head and put yourself out of your misery?" Then ran back into the formation where he had come from.

I complained to the First Sgt. about it! He thought it was funny and laughed!

I was writing letters and sending them home to my wife and my little Bubby every chance I got. Even if it was just a piece of scrap paper that I had; I would draw a picture for "My Little Bubby" and send it home.

I missed him deeply!

By October, I had still not received any mail from home and was increasingly worried as to what the situation must be back home with "My Little Bubby" and his mother.

Soon we were given "1 boiled egg" with our "1 hot dog bun" in the mornings. Then a week later, loaded on buses and taken to a fenced in area along the side of a highway. Which had unlivable mobile homes and a tent city off to the side?

It was not Army tents, it was Arabian tents.

The rumor about my being wounded by friendly fire had gone through the entire battalion. It made the First Sgt. happy when I would complain to him about it.

Since the first day, we had arrived in Saudi Arabia, we had no mission. Our duties simply consisted of pulling guard duty within our unit area inside the tent cities.

We received visits from a tanker truck that brought our water. We could never drink the water right away as it had to be tested from fear of Saudi Arabian's poisoning it.

Towards the middle of October, seemingly out of nowhere, a Staff Sgt. showed up. Who had just arrived in the country from Fort Bragg... He was assigned a cot across from me inside our tent.

The first words he ever spoke to me were, "What are you going to do when you get a divorce?" I replied, "I'm not getting a divorce!" He said again, "What are you going to do when you get a divorce?" I asked him, "What are you talking about?" He repeated himself even again, "What are you going to do when you get a divorce?"

I said to him, "If you know something that I don't know, just tell me!" He then looked me in the eye, as once again he said, "What are you going to do, when you get a divorce?

This conversation went on for three days! He would never say anything other than, "What are you going to do, when you get a divorce?"

The stress over not being able to contact my wife and not knowing if "My Little Bubby" was okay had taken its toll. Each time I had tried to call home on the satellite phone in our unit area, there was no answer! Combining that with all of the disrespect from everyone in my unit on an hourly basis was pushing me over the edge.

I was not only overstressed with worry about what was going on back home. I had also lost about 35 pounds from only having bread and water to eat for those first few months.

Then suddenly, onc day I received a letter from my wife. It said in broken English, "If someone say, they see me run round, is true. I do, porkay si, I want you feel pain."

This was it, I was now fully insane!

My thought processes were swimming in all directions!

I went to the battalion Sergeant Major and asked him to re-assign me to a different battalion!

He asked me why! I explained to him. What had happened to me since my return from being wounded during "Operation Just Cause?" He went on to say that he understood, but our Battalion needed to keep good people. He refused my request for transfer; looking back I question his motive!

An hour later, the First Sgt. came to me and said "You went to the Battalion Sergeant Major and asked for a transfer to a different battalion?"

I had just been betrayed!

I said "Yes First Sgt., I want out of this battalion as soon as possible." He said, "You sit right there! I know exactly what to do with you!"

20 minutes later he was standing in front of me saying very hatefully, "Pack your gear; you're going to Bravo Company! I've already talked to the "Bravo Company Commander, he's going to take good care of you!"

The look in his eye was that of pure evil!

I packed all my gear, walked with the First Sgt. over to the "Bravo Company Commander." There he told me what squad I would be in and pointed me into a nearby tent. Once there I met the Squad Leader and the Bravo Team Leader that I was replacing. He was very unhappy, and so was the rest of the squad. Members of the bravo team spoke up and said, "Why are they doing this? We are already a team! We came here as a team! Why won't they let us stay in a team?"

I tried to explain to them! That I had requested to be sent to a different battalion! But it was denied! Also that, I was just as unhappy as they were, about being forced to take over their team! Pushing out their team leader! For that, they hated me even more!

Well, that was it!

The stress over everything had become too much, and I clicked. It was as if, I had just come up out of that dental surgeon's chair and the "Dead Heart", he had used to spread my mouth apart, which had slipped off of my teeth and crushed into the roof of my mouth for 3 ½ hours, years before.

My thought process was not motivated by fear, but my desire to respond in the correct manner. I was thinking about every little thing that happened, considering each alternative, but ultimately the answers did not come from logical thinking. There were just too many variables and my mind was fried!

I walked down a few tents, to talk to the chaplain! There I stood waiting for the chaplain while he talked to another soldier. Over to the right side of the tent, another soldier was on his hands and knees going through different letters in a box. He found what he was looking for and left the tent. A moment later another soldier came in and started looking through the different letters that were in that box.

Still waiting for the chaplain to finish! So that I could speak to him, I was drawn to that box and that pile of letters. Once no one was looking through the letters! I walked over to it, and started looking through different letters that were there. Each one of them was addressed to "Any Soldier!"

I soon picked out one and stood up. I opened it, began to read it, still waiting for the chaplain to finish what he was doing so that I could speak to him!

After 30 more minutes of waiting for the chaplain, I gave up, and then walked back to my new units' tent. There I read the letter again and wrote a reply. That afternoon I put it into the outgoing mail. Two weeks later I received a reply. In the meantime, I had gone back to the chaplain's tent and answered other letters that were sent to "Any Soldier!"

I had also taken another turn on the satellite phone in our battalion area, and had the information operator look up a "Newspaper" in South Carolina.

After calling the "Newspaper" and getting the price for an ad that said "Lonely Soldier, Request Correspondence from single female!" I sent them a money order to pay for the ad…

Then I walked into the chaplain's tent, got a handful of letters addressed to "Any Soldier!" Searched through them until I found a single woman and started writing. Soon letters started coming in, sometimes 10 per day and saying that they were there for me, and that I could write about anything that was troubling me.

I soon started writing about sex. Using the name of the woman I was writing to. As the person, who, I was describing what sexual act I was performing with.

Very much the same as what you would read in a "Penthouse Magazine" letter.

Soon I was corresponding with 50 or more. Always the same version of the "Penthouse Magazine" type letter! But with names and small variations changed to fit the person, I thought I was writing to.

There were so many and so often. That I could not keep up with who I was writing to. Or, who I had just written to the week before!

Then one day I received a letter from two sisters, who said they were going to the community college and taking night courses. I returned their letter with the same version of the "Penthouse Magazine" type letter, but with their names and the small variations to fit into their being sisters. A few days later I received a reply, and they asked if I would send another letter.

I did and tried to take the same short story, but change it around a little.

Over the next week, I continued to correspond with the now, over 60 women never sure of who I was writing to, with more and more letters arriving daily!

Often, almost daily, when the mail came in from the states, and I received a hand full. Many people around me complained about my getting so much mail.

Then I was approached by my squad leader and ordered to give him my M-16 and bayonet. I did, and was told to follow him to the First Sgt. of Bravo Company!

Inside the First Sgt. tent, was every high ranking NCO of the company. I was then told that two sisters had received a letter from me and had taken it to their Sex ED teacher at school. The sex education teacher took the letter from them and sent it to the state governor. The state governor had then sent it to my Battalion Commander.

The same Battalion Commander, which I had served under during "Operation Just Cause" in Panama! The same Battalion Commander, who thought just leaving my "Purple Heart Medal" on my pillow while I was asleep in the hospital, was good enough for me. The same Battalion Commander, who sent down through the chain of command my United States Army orders for my 2 Bronze Stars, Expeditionary Medal with Spearhead and my Combat Infantryman's Badge. But never thought enough of me to have a company formation to read the orders and pin them on me.

The same Battalion Commander, who retracted the "Spearhead" from my "Expeditionary Medal" after my morphine induced, exaggerated value of the weapons, we confiscated out of the PDF Battalion 2000 arm's room. The same Battalion Commander, who allowed the vicious rumors about my sniper wound, to be described as friendly fire. The same Battalion Commander, that was nowhere in sight when I had been wounded in combat. On the other hand, near any of the other soldiers in our unit who had been wounded or killed!

In fact, I have no memory of him anywhere in Panama on any of the missions we went on. I had never even been within 100 meters of him. I only knew him, from the many photos posted around our company building back at Fort Bragg.

Next, the same "Bravo Company Commander," which the First Sgt., had come to me about and said "You sit right there. I know exactly what to do with you!" Then 20 minutes later he was standing in front of me saying very hatefully, "Pack your gear; you're going to Bravo Company! I've already talked to the "Bravo Company Commander, he's going to take good care of you!" was now standing in front of me, ordering me to write a full confession.

I would only write: "I wish to speak to my lawyer at this time!" which made him grab his M-16 and give me a look that said, "I want to kill you!"

Then the Battalion Commander had me standing in front of him inside his tent alone; I explained everything that had taken place and how it had gotten out of hand. He gave me a Field Grade Article 15 under the Uniform Code of Military Justice, busted me down to E-4 and took away a months' pay.

His comment after given me a Field Grade Article 15 under the Uniform Code of Military Justice, and taking away a months' pay, was, "I happen to know your marriage was broken long before we ever came over here!"

That was a startling experience to the true extent of his character!

I asked him not to take my month's pay! As I was right there next to him, and had no way of spending it. It was the only money, my wife and son had for food. No matter what he thought of me or my marriage? He refused, and as I walked away, he ordered me not to appeal, my Field Grade Article 15!

As I walked outside of his tent, off to the side, was the First Sgt., who had been told he was not needed during "Operation Just Cause." Next to him was the S-4 Lieut., who had passed out all of his equipment. So that he could have stuff his rucksack full with a pillow. Standing with them was the S-4 Ammo NCO, who was told he was not needed during Operation Just Cause. And who had worn a cast on his right fist for months after the window of my car had been busted out. Sitting down in front of them was the S-4 Section Sgt., who had betrayed me, when I told him what the S-4 Lieut. had so cowardly done.

All smiling in my direction!

I wondered if this was their entire fault! Sense they were the ones that picked up all the mail and took it to the airport. Then returned with the incoming mail!

My caution was not motivated by fear, but my desire to respond in the correct manner. I was thinking about every little thing that happened, considering each alternative. Had they opened and read my mail, then conspired against me?

As I left the area, the S-4 Captain, who I was assigned to during "Operation Just Cause" then came darting from around a tent, just in time to get in a smile as I walked by!

I was then put in the back of the Battalion Sgt. Majors Humvee, who I had asked weeks before for a transfer to another battalion (then betrayed) and driven for an hour and a half. Where I was standing in front of another 325th AIR battalion Sgt. Maj., given a different M-16 and bayonet! I then was assigned to an Airborne Infantry team, this time an Alpha-team but as a rifleman!

There finally, I and 2 other men were taken to an American housing compound. The husband and wife that were our sponsors invited us into their home for a home cooked meal. Before we ate, the 3 of us from my unit were allowed to take our first bath in Saudi Arabia. For me, it had been 3 months.

They were so very nice to us!

A week or so before Thanksgiving, while pulling a late night perimeter guard shift. I spotted a SCUD missile in flight, coming in our direction from the North. Instantly, two Patriot missiles, one behind the other, came across the sky in their intercept flight pattern, directly from the West.

As the first Patriot missile collided and exploded with the SCUD missile, the second Patriot missile kicked in its afterburners and changed directions to fly directly through the explosion.

Once it went through the explosion and did not hit anything solid, it continued on until it was out of sight!

Immediately, I picked up my radio microphone and contacted the E-6 Staff Sgt. at Battalion headquarters. He was a Ranger, and assign to Battalion CQ duty!

I said; "Salute Report Follows!"

He said with a sleepy voice; "Roger Out!"

I could not believe what I had heard. He ended the transmission, by saying "OUT!"

The correct response is; "Send Salute Report, Over!"

You never say "OUT!" until it is the very last and ending comment, meaning the entire conversation was over. Anything more, was a new and different situation.

So I said again very loudly; "Salute Report Follows, Over!'

He said again, slowly and in a half asleep voice; "Roger Out!

By this time, the explosion was all fluorescent pink smoke, and I knew the Staff Sgt. "Ranger" was asleep on duty at the other end of the radio. So I yelled, "Emergency Salute Report Follows How Copy, Over?"

I did this in hope that he would wake up, because I desperately needed to send up this information. Again, he says in a sleepy voice; "Roger Out!

The pink fluorescent cloud of smoke was now even larger, so I donned my chemical protection mask and yelled as loud as I could "Gas Gas Gas!"

Someone came out of a tent in the darkness and looked in my direction. So I repeated myself while using the hand and arm signals, and yelling "Gas Gas Gas!"

He ran into his tent and came out a moment later, with his gas mask on and started repeating my message "Gas Gas Gas!"

Moments later, I could hear yelling from further away; the message was being relayed from soldier to soldier "Gas Gas Gas!"

Within one minute the entire battalion had on their gas masks.

At that point, I tried again on the radio with the Ranger sleeping at his post, and yelled into the microphone; "Gas Gas Gas, Emergency Salute Report Follows, a SCUD Missile

Explosion In Air, One-Mile to the North, Pink Fluorescent Smoke Cloud! How Copy Over?"

Again I heard his half asleep reply; "Roger Out!"

All I could do was shake my head from side to side... As I thought of my DHG, (Distinguished Honor Graduate) of the NBC (Nuclear Biological Chemical) Warfare School, and what good it did! When the person on the other end of the radio, was asleep at his post!

Chapter 5

DRIVER

During the 2nd week of November 1990, our platoon leader and platoon sergeant called formation. Once everyone was formed up, the platoon Sgt., said, "We have one of the new deuce and a half truck's assigned to us. We need a volunteer to drive this truck!"

I remained silent and so did everyone else! The platoon sergeant said, "Now come on, we need a volunteer, somebody speak up!" Again, everyone was silent!

I had been a driver for the Battalion XO in the 505th years before. I had driven dump trucks and semi-tractor trailers before I came into the army. I knew the extra hours in the motor pool, and of the extra work that comes along with being a driver of a military vehicle. So I did not speak up; let somebody else have this!

Five minutes later, still no one volunteered! So I said, "Ok Platoon Sgt., I'll do it!" Shocked, and with his eyes opened wide. He said very surprised, "You!" I replied "Yes, Platoon Sgt., I'll volunteer!"

I spent the next couple of days, with the driver that brought us the truck; he was from a National Guard unit.

Soon the day came, that we were to relocate to the border of Iraq! The Battalion was to be flown, and then bused to the location on the border with Iraq.

All trucks would be in a convoy, catching up to our unit within 48 hours. Once the deuce and a half pulled up with the national guardsmen, of which I was to drive. My platoon leader walked up to him and said, "If you let Cloak drive, I will ensure that you receive a Field Grade Article 15!"

Surprised the driver said, "What?" Again he said, "If you let Cloak drive, I will insure that you get a Field Grade Article 15."

I wasn't sure what was going on. This did not make sense! They had asked for a volunteer, no one would volunteer, so I spoke up.

The driver and the Lieut. soon disappeared behind the tents, as I waited.

Once the national guardsmen returned, he said "Okay! Get in the truck. Let's get out of here; so I loaded the truck on the passenger side, and we left the unit area and found our place within the convoy of trucks that would be headed north to the border of Iraq.

Six hours into the trip, I spoke up and asked the national guardsmen; "When do you want me to take over driving so you can get some sleep?" He replied, "You're not driving!" I said "Sure I am! I'm supposed to help you drive, during the trip to Iraq!" He spoke up as his eyes opened wide and said "No, you're not driving!" I said "You mean you're going to drive the entire 48 hours?" He said "Yes!" I said, "That's not a good idea, you need to get some sleep." He replied, sternly; "You will not drive this truck!"

I did not want to get him into any kind of trouble, although I also knew that no one would know whether or not I drove that truck, unless he told them.

I did not know this man, I also did not know what he had been told when he left with the Lieut. and the platoon Sgt. I also knew that when they asked for a volunteer, "No one would volunteer!" So I spoke up and said, "I volunteer!" Once it was me, the situation changed, and they did not need a volunteer.

It was obvious that I was still being punished, or somebody had some harebrained idea about what I might do with the truck. I just don't know, why, if anything they were thinking at all.

My thoughts were not motivated by fear, but my desire to respond in the correct manner to the situation at hand. I was thinking about every little thing that happened, considering each possibility, but ultimately the answers didn't come; there was just no reason for this kind of behavior!

Once we made it to the border of Iraq, the driver parked the truck one half mile from my unit. I spoke up and said, "You might drive up a bit closer, where I don't have to carry so much of my equipment that far.

He said firmly, "NO!"

Again, I spoke up and said "You could just pull right up to it, and I can get out, then you could drive back!" He looked me in the eye and said sternly, "No, you will walk!"

While I was taking down my equipment off the back of the deuce and a half truck, my team leader walked up and said "Cloak, let's go!" I replied, "Okay! Just one moment, I've got to get my duffel bag down from the back of the truck."

I had kept a canned ham that my aunt had sent me and some cookies from my mother. I was hoping to share them with friends on Thanksgiving Day! That compared with the many items of equipment made it very heavy!

I did not get to share it on Thanksgiving Day, but did later in the year, on Christmas morning. Still though, no one said "Thank you!"

As I put on my rucksack and picked up my duffel bag, I said "Lead the way Sgt.", he took off walking, and I attempted to walk alongside him, but soon he was out far ahead of me. As my duffel bag was slowing me down because of it being heavy and my needing to drag it along the ground.

When I arrived at the foxhole I was assigned to! My team leader had already started digging it and had already finished digging his sleep hole.

Immediately, he pointed to a spot right next to his sleep hole and said "Dig your sleep hole right there!", so I did!

Chapter 6

The Foxhole

When my sleep hole was dug, I walked over to the foxhole my team leader was digging. There I sat down on the ground disgusted and said, "Ranger, do you realize where you are?" He looked up at me and said, "What do you mean!" I said, "Look at the terrain, we are in a crater! Once you dig the foxhole down to an armpit level, and we are standing down in it, you will not be able to see very far in front of you. The outer edge of the crater is going to block your view!"

He looked at me and says "Cloak, the Lieut. said to dig the foxhole right here!" I replied "Well, if we get attacked, we will not be able to see anybody coming! So this whole fighting position is inoperable! There's no way, once you're down inside the foxhole at armpit level, that you can see more than 5 meters! Look at the foxholes, off down into the distance, going out of sight! You can see that every one of them at an armpit level; are on a level with the terrain for miles? You're a Ranger; didn't they teach you that Terrain Dictates?"

He stopped what he was doing, looked at me, then looked down at the foxhole and says, "The LT said to dig here, and I did!" I said "Well, we need to move it up 5 meters to the edge of the crater." He said "No, the Lieut. said that we are to dig it right here; so that's what we have to do! We have to dig it right here, where the Lieut. said to!"

"Then we need to speak to the Lieut., I said, because he knows that terrain dictates or should know! He was probably in a hurry and was only thinking of what he read in a book about how to put each foxhole 50 meter's apart. Everyone with experience knows that 50 meters are only a guide line, and that adjustments have to be made, if the terrain makes the location of the foxhole useless; and here, in this crater, once down at an armpit level, you cannot see but a few meters. So moving the foxhole up or over a few meters, makes it useful! Which will allow us to see an enemy long before they reach us? If not, we cannot see them until they are close enough to through a grenade!"

Soon, the Lieut. was walking by so I spoke up and said "Lieut. could I speak to you for a moment?" Disgustingly, he says, "What is it, Cloak?" I said, "Lieut. Sir, this foxhole is down in a crater. Once we dig it to an armpit level and get down into it. We cannot see more than a few meters in front of the foxhole. That makes this foxhole inoperable."

Wow he was so angry! The Lieut. says, "Cloak, I stepped off this lazy W, for every foxhole in our unit. The measurements put this foxhole right here." I said "Yes Lieut.,

I understand that, but in the book you read, when putting up a perimeter of foxholes, and you step out your lazy W., where the foxhole is every 50 meters, terrain dictates whether or not to move the position, right or left, forward or back a few meters.

He interrupted me and says "Cloak, the foxhole is going right here." I spoke up and said "Lieut. Sir, with all due respect, if you'll just get down here for a minute into this foxhole, you will see what I'm talking about. I can only see a few meters in front of me. We need to move this position a few meters to the edge of the crater. And, dig down to an arm pit level, so that we are looking over the edge of the crater instead of looking at the outer wall of the crater. Only then, can we be useful to the fighting positions to our left and the fighting positions to our right if we are attacked!"

"If not, and we are attacked! We have to stay down inside the crater, waiting for the enemy to get so close, that they could throw a grenade. And, because we're down in the very bottom of this crater, a grenade just keeps rolling and rolling towards the bottom, which is where we are!"

The Lieut. Spoke up and says "Cloak, I don't want to hear anything more about it. This is where the foxhole is going to be dug, and you're going to dig it." He then left!

Disgustingly I turned towards the Ranger and said "The only way we can make this foxhole of any use, is to not dig any deeper than we are right now. What we have to do is, build up the sides of the walls. Up to where they are at armpit level. This is, as you can see, with me standing in the bottom of this foxhole, as deep as we dug it. Right now, this is at an armpit level. Pointing my arm straight out as if it were my M-

16, just barely points over the top of thc outer edge of the crater.

It's only from the standing position I'm in right now, that we could shoot at anyone approaching us if they are an enemy. And, they are more than 5 meters from us. If not, and we keep digging down further, and we get down in that foxhole during the fight. We cannot see anybody more than 5 meters from us, which means we will have to get out of the foxhole and walk to the edge of the crater. Then use the front edge of the crater, while we lie down and return fire. Only then will we be able to see who is really there!"

We spent the next week building the outer edge of that foxhole up with small rocks; we could find lying around in the desert, and then pushed sand up to it trying to fill up the area inside that crater immediately around our foxhole.

Then one day, the Lieut. walked by and says, "What are you doing Cloak?"

I explained!

This totally ticked him off; He ordered us to stop, so we did!

I had run into this type of stupidity before. A Lieut. would spend a few weeks' reading books. Then they would go to "Airborne School" where they would get 5 jumps.

Then they come to a unit inside the 82nd Airborne and refused to listen to anybody that had more experience than they had. Because the person explaining the situation to them, had a lower rank than they had; and it didn't matter if it was an E-9 with 20 years experience or not!

I was clearly an "Unwanted, Unappreciated Combat Wounded Veteran with 7-years of military experiences against his 1 year!"

The small hand held radio I received in a box sent from home, which my mother had put in it for me; had a set of small earphones. I could not get it to work; it would not pick up anything!

I walked over to my sleep hole and stepped up on the pile of dirt beside of it. I put the small radio in my right hand and starched it up toward the heavens as far as I could! It suddenly almost picked up a sound!

As I turned my hand from side to side, trying to catch the radio waves of "Radio Free Europe." The sound of "George Strait" singing "Love Without End, Amen;" came through and filled my eyes with tears! I missed my little "Bubby" so very badly!

4 weeks later, we were allowed, one by one, to go into the only Army tent near our location and take a standing bath next to a bucket of water. That was the second and last bath; I was allowed to take during "Operation Desert Storm and Operation Desert shield!"

The only Lieut., I had ever worked under, who did not act this way, did graduate from "West Point!" He said, "When the shit starts rolling down hill, it stops on me! I will not let it find its way to you!"

And he never did!

Chapter 7

Chop to the Throat

Beautifully absent of any clouds, bright and blinking stars brilliantly filled the clear night sky.

Even though there was no wind, the bitter cold of the 17-degree winter air, cut through my uniforms, as I had put on a double pair, and chilled me to the bone.

Two pairs of uniform, was not enough, over whelmed by the freezing cold night air, I further, put on my chemical protection suit! With its caracole lining, it proved to do the trick. Although, it left me burdened and unable to move quickly!

Looking through the PVS-7, night vision devises, which intensifies any star light, a billion more stars filled the heavens. Much more than I had seen during times past!

Doing a tour of guard duty, at night, all alone, is a dreadfully boring experience. The knowledge that, the enemy is 3 miles away isn't all that worrisome.

However, experience reminds you, it is highly possible the enemy has already sent out a platoon size group of their specially trained soldiers. On a search and destroy mission! All Hell, can beak lose at any given moment! Performing your duties within your most professional attitude and abilities is a life saving necessity!

The existence of those sleeping near you, inside your guard duty patrol area, rest entirely on your professional performance!

It was the middle of February 1991, at 2:30 AM; I was ending my four-hour tour of guard duty, on the southern border of Iraq while everyone in my unit was sleeping.

The frozen desert sand cracked and crunched under my body weight, as I walked around checking each platoon member for signs of life. While erratically scanning the surrounding area with the night vision device, to insure, we were not attacked without warning. That warning rested entirely on the person or person's on guard duty, me!

I was to wake up E-4 Robson <altered name> so that he had time to climb out of his sleeping bag, put on his boots, standup and be fully alert for his guard duty. Which started at 3:00 AM and would last until 5:30 AM? This being, the only time I would have to sleep for the next 24 hours.

Robson had completed Basic Training, Advanced Individual Training in the US Army Infantryman's Military Occupational Status and Airborne School. He was now serving during his 2nd year.

Everyone in our platoon below our platoon sergeant was afraid of him. For the reason that, he was around 6'3" tall, 225 pounds, and a golden glove boxer!

He could do no wrong!

Prior to being assigned, to the 82nd Airborne! I had completed 9 nonstop months of Basic Military Training, Advanced Individual Training in the US Army Infantryman's Military Occupational Status, Airborne School, Special Forces Qualification Course pre-faze, Special Forces Qualification Course 1st Faze, which consisted of Advanced Hand to Hand Combat qualification, Long Range Patrolling Qualification, Wilderness Survival Qualification, and Expert Land Navigation Qualification among others.

Second Faze Weapons, which was qualification with every military light arms' weapon in use throughout the world at that time, from old, to super modern. I spent the 3 years of my enlistment at Fort Bragg, then went on to serve 2 ½ years of re-enlistment in the Panama Canal Zone leading up to "Operation Just Cause," as a team leader.

During that time, I had met many soldiers who fit the description of his charter. They were the tallest in the group, strongly built, but unfortunately obsessed by the knowledge that no one smaller than they were, had ever opposed them. And rightfully so, for they were the best there has ever been! No one fit, in comparison to his greatness as, The Chosen One!

My land navigation skills, which were thought to me by the Green Beret's in Special Forces school, were better than anyone in that entire airborne infantry battalion...

Which caused animosity among everyone who tried to outperform me while land navigating through the triple canopy jungles. Simple because, I always ended up within 5 to 10 meters of the destination!

Thanks to the skills taught to me by the Green Beret's, I never failed; they failed 9 out of 10 tries and never thought to ask how I could do it so easily! Nor had anyone ever thought to look in my records, for school and training accomplishments.

Their idea was to take away my map, at night, and only allow me to have a compass. Then tell me to take them to the small house 3000 meters away, at a 120 degree direction.

I never failed, and they always hated me for it!

Before "Operation Just Cause," I had gone to NBC School. I graduated as "Distinguished Honor Grad." Graduated BLDC (Basic Leadership Development Course), BNCOC (Basic Non-Commissioned Officer Course), JMS (Jump Master School) which puts a "Silver Star" above your jump wings and makes you a senior parachutist.

I had also received a Bronze Star for parachuting into combat during "Operation Just Cause," CIB (Combat Infantryman's Badge) a Bronze Star for Valor in the Face of an Armed Enemy and Purple Heart for my sniper wound. I also, had worked a year as an Air Item NCO, in the S-4 Section of the 325 A.I.R. Headquarters' of Headquarters' Company!

I had no fear of Robson; in fact, I had a little animosity toward him. It had built up a week before, when at night, our platoon had been sent out on long range patrol with Robson as point man/lead men. He had gotten us lost. Not until many hours later, had another platoon crossed our path, before we again knew our precise location on the map.

Many times, someone spoke up and said, "Let Cloak do it! He has more experience than anyone here!" This set Robson's ego on fire! Firmly, imbedding in his mind that I was going to have to be proved worthless! Ether this or he would lose his status as point man.

Whispering to our squad leader in the dark, he says, "Alright let Cloak take over as point man! You'll see, he'll get us lost!"

I heard this and walked over to them. "Where do you want me to lead you?" I stated confidently!

2000 meters at a 60 degree heading, is a 50 gallon drum that headquarters platoon put out as a land marker! We were supposed to have been there 3 hours ago, for a radio communication check with higher! Can you get us there, asked the squad leader?

Derogative, I replied, "That's too easy!"

Within a very short time, we were at the drum and finished with our late radio check with higher! Robson was shocked at how easily; I had gotten us there in the dark and stated, "You were lucky this time, Clock!"

At that precise moment, I knew that we would soon confront each other. Out of vindictive spite, through eye's slanted in cold heartless dislike. I replied, "Yeah, right!"

The only other man in our platoon, that was the same height as Robson, was Johnson <altered name>! He was a nice enough fellow! Just one flaw though, he also feared Robson, to the extent that he was going to follow Robson as his sidekick, ensuring that he never had to stand up to Robson.

It was now, time to wake up Robson and get my turn at sleeping! At five minutes to three o'clock, I stood up and grabbed the night vision devices. That I was to give to Robson to use during his shift on guard duty!

Each of us in my unit was assigned to a foxhole and directly behind that foxhole, we had to dig a sleep hole, about 1 to 2 feet deep. The length of our body to include a space for our backpack!

I knew the location of Robson's foxhole and easily found his and Johnson's sleep holes, as I walked up on the area in the dark of night. Once there, I noticed that the sleep hole of Robson's and Johnson's was empty.

Looking around me across the sand, shining in the starlight, the nearest sleep hole, which was our squad leader's, had a shape of a body on its side, sleeping. Robson and Johnson were nowhere in sight.

The next foxhole to the north or the south was 50 meters, in that direction. Directly behind them, were the sleep holes. Because there were only two unit members assigned to each foxhole, this meant that they would only have one sleep hole each, 2 in total.

While pulling my guard duty, where I had positioned myself up on a small mound! It was, maybe 4 feet higher, than anywhere else in the area! I had not noticed any movement in our entire unit's perimeter.

As I stood there next to Robson's empty sleep hole. I looked through the night vision devices and turned in a 360° turn (A Complete Circle!) while watching for any sign of movement. There was none!

The temperature felt like it was below 15 degrees now. As cold as it was, I wanted to find Robson and get him to wake up. I needed him to wake up, so I could give him the night vision device. Only then could I go to lie down in my sleep hole that was over 200 meters away.

I was tired and past ready to sleep!

There in front of me, 30 meters away from Robson's sleep hole, was the dark outline of a foxhole that had no roof. It was a black hole in the darkness.

Walking up to the foxhole, and standing in front of it. I called out, Robson, Robson!

There was no reply, so I stated again, "Robson, it is me, Cloak! It is time to wake up, your turn for guard duty!"

Still, I heard no reply!

Not wanting to raise my voice, which would carry a long distance across the desert, I squatted down on my knees by the foxhole. Putting both my hands on each side of my mouth to channel the sound of my voice, I said again, and more than loud enough to wake somebody up. "Robson, Robson, Robson, it's time to wake up. It's your turn for guard duty." Again, nothing but silence followed my voice!

As I leaned forward trying to focus my eyes, I could not pick out a silhouette down in the foxhole. All I could see was blackness!

I put on the night vision device. Turned on the switch, which turns on the infrared light, lean forward as far as I could without falling into the foxhole. Then peered down inside!

I could see nothing but the lower far side of the foxhole. It was too deep to see to the bottom at that angle.

Instantly, I realized that as tall as Robson was! Which was around 6 foot 3? His foxhole would be more than too deep for me to bend over and look in and be able to see into it directly, straight down to the bottom without pushing in dirt and rocks onto the person or person's down there?

Again, I place my hands on both sides of my mouth and cup them to channel my voice and spoke louder saying repeatedly, "Robson, Robson, Robson!" Continually getting louder and louder and louder! For I knew he must be down in that foxhole! There was nowhere else for him to be!

Worsening the situation, his foxhole buddy was down there with him, obviously ignoring my calling out to them!

An image soon visualized in my mind: I could picture Robson and his foxhole buddy. Both well over 6 feet tall. Lying way down at the bottom of this foxhole curled up in the darkness, with a grin on their faces.

I continued to call repeatedly, "Robson, Robson, Robson!"

After about ten minutes of this, I stood up and said, "Robson, it is past your turn for guard duty, go ahead and get up. I know you hear me, just get up. I'll be back in a few minutes to give you the night vision device."

About 5 minutes later, as I walked back towards Robson's sleep hole. I assumed he would go ahead and get on up, and would get the night vision device by walking over to me. Which would ensure that he was fully awake, allowing me to relax and get some sleep.

I stood by his sleep hole for more than five minutes waiting for him.

He did not get up!

It was now easily 3:20 AM, and far too long into my sleep time than I wanted to endure.

Walking back to the foxhole once again! I got down on my hands and knees, leaned over as best I could without falling into the foxhole. All over again, I began repeating as if I were a broken record calling out, "Robson it is time to get up, Robson it's me, Cloak! It is time to get up. Robson, Robson, Robson wake up!"

By 25 minutes after 3:00 AM, I was becoming very angry. I knew in my heart, he could hear me. He knew it was his turn for guard duty; he was fully awake and just lying there in the darkness at the bottom of that foxhole. Listening to me repeat myself, time and again, and even, time and time again! While he and his foxhole buddy smiled and laughed to themselves. Flattering themselves, content with their betrayal!

For a few minutes I sat there, quietly. Holding my temper, holding it back and holding it back and holding it back! My ego enraged with desire….

At this point, I knew absolutely! He was wide awake, although I had heard no movement or seen any movement. I knew he was in that foxhole, and he had heard every word I had been saying.

So I started talking to him. He was wide awake, and I felt it along with visualizing his face in my mind, saying, "Robson, I'll put the night vision device right here. When you come up out of the foxhole, you can find it easily. Do you need any water for your canteen before I leave? I've got a little extra if you want it! So, that's a pretty deep foxhole you're in, I mean, I can't even see you! You are way down in the ground! All I see is pitch darkness!

Still, I heard no movement, no replies to altogether my recurrent callings!

At 30 minutes after 3:00 AM I stood up totally angered, and said, "Robson get up! Robson get up!"

While getting evermore louder and louder, I was not going to tolerate this any longer. I would keep getting louder until I woke up the dead. I knew he could hear me. I knew he was awake, and this disrespect had gone absolutely totally and completely too far.

Again, I said "Robson" and then getting louder, "Robson" and then getting even louder "Robson" and still I kept getting even louder as I called out his name, "Robson!"

Suddenly, I heard his voice as he said "Cloak!" If you say one more word, I'm going to kick your ass!

I knew he did not want me to wake up everyone and let them know that he was hiding when it was his turn to pull guard duty.

Immediately, I said, "Robson, I have been standing here calling you for over a half hour, it's your turn for guard duty!"

Now it was Robson, repeating himself. He stated, "Cloak, if you say another word. I'm going to kick your ass!"

Again, I said "Robson, I have been standing up here calling you for the last 30 minutes, and I know you heard me calling you, now get on up! Get up and come get the night vision device, so I can go get some sleep!"

Feeling my heart beat next to my ear drums, he stated, "Cloak, I told you! If you say one more word, I'm going to kick your ass!"

My eye lids slowly closed then reopened! I clicked, and said loudly "Alright Robson, kick my ass! However, you're going to pull your guard duty."

Repeating himself even again, energized the savage beast within me as he stated, "Cloak, I told you! If you say one more word, I'm going to kick your ass!"

With a smirking smile, and eye's widely open, once again, I repeated myself even more derogative saying, "And Robson, I said kick my ass! However, you will pull your guard duty."

Immediately, I heard movement at the bottom of the foxhole as Robson stood and his face came out into the starlight.

I took one step back then 2 steps, to allow him up, and waited as he climbed up out of the foxhole.

Defiantly, when he had climbed completely out of the foxhole! I moved my hand with the night vision device toward him and declared, "It is your turn for guard duty!"

Towering over me with his head bent down looking into my eyes. He says, "Cloak, I told you! If you say one more word, I'm going to kick your ass!"

With my head bent back looking nearly straight up. I said to him in a commanding voice, "And Robson, I said kick my ass! However, you will pull your guard duty."

At that instant, his right fist came up to my left cheek. Then his left fist came up and hit my right cheek. Then back to his right fist repeatedly as my head bounced side to side from his pounding!

My cheeks burned from the blurry of punches as I took two steps back and said defiantly, and very derogative, "Are you awake now Robson!" The whole time as I kept my right hand on the stalk of my M-16A2, and my left hand around the barrel hand/heat guard!

Instantly, he stepped up to me lifting his right foot high above my right foot, pulled his right hand way back into a cocking motion to chop my wind pipe. Just as instantly using his full body weight sweeping his right foot back against my right foot!

Thus, sweeping my right foot out from under me as he simultaneously chopped my wind pipe, thumb first!

I had learned this tactic during my "Hand to Hand Combat" training in Special Forces school! It was used to kill an enemy, by breaking their neck! I had also learned the counter tactic of tightening my neck muscles to take the blow.

It worked, and I went flying through the air about 5 feet off the ground to land hard on my back many feet away. The whole time keeping a solid grip on my M-16A2 rifle!

Burdened by the many thick layers of clothing, I rolled to the right; placing the butt stock of my M-16A2 rifle on the ground beside me, and used it to help me get to my feet. Looking at them standing shoulder to shoulder, Robson's foxhole buddy was now joining in.

Simultaneously, they both walked toward me, as I lunged forward with my left foot first and jammed the rifle barrel into the left rib cage of Robson, just below his heart.

The same way we had trained to use our bayonet to kill enemy, causing a rush of air to leave his lungs and a stunned look to over-take both their eyes!

Robson and I locked eyes, and then Robson grabbed the rifle barrel of my M16A2 and tried to move it to the left. This filled me with joy, as the adrenalin rushing throughout my veins felt fantastically energizing.

I jerked it back toward me! Then jammed it back into his rib cage just under his heart, and click the safety switch off of safety, onto 3 round burst! At the same time putting my trigger finger in place...

Again we locked eyes! This time he did not try to move my M16, but kept his grip on the barrel. We stared into each other's eyes wanting to kill each other.

We were at a cross road, standing together! One would live, one would die, and the odds were in my favor.

A few silent seconds went by; I felt good, I felt alive and in total control. In just under one heart beat, I knew that I did not want to go through the rest of my life with the knowledge that I had killed a fellow soldier.

Confident that I had stood my ground and proved my point! I slowly, eased backward as he let go of the barrel.

Once I was a foot further away, I lowered my weapon and said sternly, "Are you awake now Robson?" He did not answer, so I turned to the right and leaned down to pick up the night vision device that was on the ground near us!

Immediately, as I started to stand, Robson's foxhole buddy grabbed me from behind. Left hand on my jacket collar, right hand under and between my legs, lifting me high in the air over his head! Then slamming me on the ground while moaning like a spoiled child that was breaking something during a temper tantrum!

I rose up with my weapon pointed at them both, as they stood there together facing me.

The content of their character was very visible in the starlight! They were 2 peas in a pod.

Two and two do equal four!

Although I had been protecting them while they slept with my very own life, they hated me to the extent of wanting to see me dead!

The night vision device had been thrown through the air, during all of this. It had landed many feet away!

Turning back to the left, I walked over to the foxhole where the whole problem started and picked up the night vision device from the ground. Then walked over and put it on the ground in front of the 2 of them, turned to walk away, took 3 steps back toward my foxhole and sleep hole 200 meters away!

There I could clearly see in the starlight about 30 feet in front of me, our squad leader, a Staff Sergeant! He was lying in his sleep hole and had himself prompt up on his right elbow looking straight at us. He had been there the whole time watching.

The content of his character was shocking!

I did not know him very well; he had only arrived in the country a week before. He was from a National Guard unit, assigned to our unit and put in command of us during the middle of Feb 1991; I had been in country sense Aug 9th 1990, was more experienced and more qualified than he was!

The next morning we were up at 5:30am doing calisthenics until 6:30am. All members of our squad were questioning why my face was swollen and red.

I gave them a short version of what happened! Which was simply? "Robson punched me out!"

An hour later, I was summoned to the "Platoon Leaders Position" and standing at attention in front of our "Platoon Sergeant."

He said, "Cloak, don't kill anybody!"

I could not believe what I was hearing and said "Excuse me, Platoon Sergeant!"

What I was hearing didn't make any sense! "If I had wanted to kill anyone, it would have taken place long ago!"

He said again, even more sternly this time, "Cloak, don't kill anybody!"

A scenario formed in my mind of Robson and his foxhole buddy going before the "Platoon Sergeant" and telling a completely different version of what took place. It was painfully obvious; the "Platoon Sergeant" believed every word of it! This meant anything I had to say, mattered not!

I stood silently for a moment and then said "Yes, Platoon Sergeant!"

As I walked back to my foxhole, my team leader questioned me as to what took place, so I told him!

He had nothing to say!

A few days went by, and while sitting among three other team members I attempted to make a joke! Immediately, one young man, around 18 years of age in our team, said "Cloak, if you don't shut up, I'm going to kick your ass!"

Shocked I said, "What?"

He repeated himself saying, "If you don't shut up. I'm going to kick your ass!"

There was no question in my mind, whether or not I could defend myself! The question was, how deeply ingrained in my unit was this camaraderie with Robson!

I would need to sleep with one eye open!

A few nights later while sleeping, I woke from a horrific nightmare. In this nightmare, I was in the kitchen of a home I grew up in. When I heard someone say "Bubby is drowning, Bubby is drowning!" I walked outside and looked toward the creek that was running by our home. As I looked in horror in that direction trying to focus and find my little Bubby a voice said, "It's okay, your brother saved him!" I then turned and walked back into the kitchen. As I looked towards the sink a nightmarish feeling engulfed me!

At that moment up through the opening of the drain came squishing out what looked like rotten packed macaroni! Instantly, I had a length of thick wire, much like a clothes hanger straightened out!

As I sensed evil, approaching from behind me, I turned and tried to hit it with the piece of wire. I just missed it! Then again, I felt it approaching from behind me and speedily turned in the opposite direction and whipped out with the wire, narrowly missing it. Repeatedly, each time I sensed it, I would turn and strike at it trying intently to catch this evil and hurt it.

When suddenly a voice said, "You have done well!" Then the dream ended!

Later in my life after leaving the military this dream would unfold in my natural life, partly being the spouse and her activities that you read about at the beginning of this book. And, what you will read in the pages to come! So take note!

Stress was building in me from not being able to sleep, because while sleeping every time someone walked by my sleep hole during the night, I became instantly widely awake, watching, every move!

It had gotten to the point where I realized, I was not going home until we went into Iraq and the war was over!

So that night while lying in my sleep hole, silently inside my mind, I concentrated intently and prayed by the name of "Jesus Christ of Nazareth."

"Please, father God! It has become very obvious that I am not going home to my little Bubby, until after I go into Iraq and this war is over!

Please, father God, by the name of Jesus Christ and all that is Holy, allow the ground war to start within the next two weeks and not last more than three days!

Please, father God! Don't let anyone with me get killed or hurt this time, like, they did during "Operation Just Cause!" Then father God, please allow me to go home as soon as possible and find my little Bubby, hug him and kiss him, and never let him go!

Please father God, allow me to watch him grow up safely in the same location that I grew up in, and please, father God, allow me to find a karate teacher, that will help my little Bubby learn to defend himself.

Please, father God, allow me to live long enough to see my little Bubby graduate from college and beyond. Amen and Amen!

A week later while packing my rucksack for a night mission into Iraq! My team leader, a ranger, walked over to me with a few items and said, "Cloak, put this in your rucksack, I don't have room for it in mine! So I did just that without question!

After a few minutes, he showed up again and said "Cloak, put these items in your rucksack, nobody has room for it!

Again, after a few more minutes he walks up with an arm load of different items such as a spool of wire, extra radio batteries, and so on.

My rucksack was already full, so I said, "You need to find somebody else to help carry this stuff! I'm already so overloaded I can't pick it up. Look at it, try yourself, go ahead and pick it up."

He said "Cloak, you put this into your rucksack and carry it. I don't want to hear another word about it!"

As I looked into his eyes I understood fully! "I was to be punished by carrying more weight than anyone else!"

Was this just coincidence, or had he been in contact with the Lieutenant from my old unit? Who had given out all of his equipment to me and other soldiers to carry into combat, so that he could fill up his rucksack with a pillow and run like a rabbit if needed?

Alternatively, was it two plus two, equaling four; my cross to bear?

I filled my rucksack (Backpack) to over filling and then tied everything else on the outside!

Our platoon was forming up for a mission briefing; my team leader had already left for it. My rucksack was so heavy I could not pick it up! I stood there angry, looking in the direction, we're everyone else was forming and laying down their rucksacks.

I became increasingly angrier as the minutes went by! Finally, my team leader walked halfway back to me and started calling "Cloak, pick it up, let's go!" So I called back

"It's too heavy, I cannot pick it up! You need to give some of this equipment to someone else. You might want to take some of it yourself. There's enough here for 3 people!"

He yelled out, "You pick that rucksack up and put it on! I don't want to hear another word out of you! Now come on!"

I was at the point where I wanted to cuss out my team leader! However, if I did, I would lose another month's pay, and another demotion in rank!

The more I stood there and watched him walk away the angrier I became! Until I rolled the rucksack over on its back with the frame side up, lay down on it and put my arms through the straps, tighten them down as tight as they would go.

I then used my M-16 as a crutch as I rolled to the right, put the M-16A2 stock into the sand, then put my hands up on the front sight and pulled myself up on one knee. Then up to a squatting position. Then slowly standing up and walking over to where everyone else had put down their rucksack!

Everyone was gone to the mission briefing, except my team leader. As I walked up, he said "Cloak, you stay here and watch everyone's equipment. I did and he left."

About an hour and a half later, everyone returned and started putting on their rucksacks. I laid down on my rucksack put my arms through the straps and use my M-16 as a crutch to help me slowly stand up.

Without any briefing at all, I got in line in my position within the team and followed as the entire squad headed off at a fast pace in a northern direction.

I could not keep up; the rucksack had to be close to 300 pounds. All I could do was lean forward and then keep my feet under me as the weight of the rucksack carried me forward.

Soon I was far behind everyone in the squad.

I had not been given any kind of briefing as to our mission. So I had no idea where we were going, or why we were going there. I didn't even have a map!

Up front of me was our squad, which was running off and leaving me. I was falling further and further behind. Soon they would be out of sight!

We soon came to a rocky area that must've been easily 1000 meters across, with big rocks the size of grapefruit's scattered so thickly that I could not take a step without stepping on top of one of these rocks, and it trying to roll under my foot.

By now, the rest of the squad was a good 500 meter's ahead of me. And not slowing down! Suddenly, I heard someone say "Cloak!" from behind me. I turned to look!

There behind me, about 100 meters was another team member. His rucksack was also overloaded. I had not noticed him before. So I slowed down and waited for him to catch up to me, and we walked the rest of the way together until we came up on our squad as they put up a hasty ambush in a valley, inside Iraq.

We spent the next hour digging fighting positions. Stretching communication wire to each position, and setting up claymore mines, to the front, in the kill zone.

I was commanded to a position next to the squad leader. As nightfall came upon us, he said "Cloak! You take the radio; I'm going to get some sleep, wake me up, if anything happens."

The night was uneventful, up until one o-clock a.m.; at which time I woke up the squad leader and informed him that I was falling asleep, and that he needed to wake up and take over while I had a turn sleeping. After about 10 minutes, he agreed and sat on up.

About 10 minutes later, I could hear him whispering into the radio microphone, but could not make out what he was saying. A couple of minutes went by, and he whispered into the microphone again then became very silent.

Suddenly, without warning! Hundreds of short range missiles began firing, off from the backs of trucks. Right over our heads from maybe 150 meters to our south, going into Iraq!

They were lined up East to West's as far as the eye could see, these missiles light up the night sky nearly given me a stroke, heart attack combination!

After about five minutes of this continuous firing of missiles, it suddenly stopped. After about 15 seconds, I turned to my squad leader and yelled sternly angrily, "You Could Have Warned Me!"

The next morning was the same as the day before. I and the other squad member who had been exceptionally overloaded with other member's equipment, arrived 30 minutes after the rest of our squad, back at our foxholes.

We were ordered to take our E-tools and fill in our foxholes. Formation in 3 hours! We did, and soon a deuce and a half Army truck driven by the National Guard member arrived at our platoon area. The same man and truck that I had ridden with coming to this location!

We were all ordered to pick up our rucksacks and tie them using the shoulder straps on the outside of the deuce and a half truck.

Our entire platoon was to get on the back of this one truck. It was impossible for everyone to tie their rucksacks all at the same time on the side of the truck.

So the platoon leader, a lieutenant, picked 4 people to tie all the rucksack's, slung over the side of the truck. Then ordered everyone else into the back of the truck!

I was one of those people ordered into the back of the truck. Leaving my rucksack lying on the ground! Someone else would have to pick it up, and tie it onto the side of the truck.

As I watched, one soldier picked up a rucksack, held it up to the side of the large truck. Another person in the truck crabbed it, put the straps through the boards connecting the straps back together and tighten them down.

This went on until there were only two rucksacks left. One mine, the other rucksack belonged to the other soldier who had been overloaded also.

You might say we were too overburdened infantryman.

Then, when one soldier tried to pick up my rucksack and could not, he reached for the other rucksack. He could not pick it up ether, and asked for someone to come help him! One of the people that had already been helping him came over and the two of them strained to pick up my comrades' rucksack!

Then the two of them took it over to the truck and held it up while it was strapped on. They then walked over to my rucksack and attempted to pick it up. They could not, and stood around complaining about how heavy it was. Claiming that the two of them could not pick it up!

The platoon leader, our lieutenant, noticed this! Climb down, from the front of the truck. And said, "Pickup that rucksack, let's go!" Both soldiers looked at him and said simultaneously, "It's too heavy, we can't pick it up!" The lieutenant became very angry and said "I told you to pick it up!" At that point, the two soldiers, both at the same time, bent down and grabbed a hold of it with their hands and attempted to lift it off the ground.

They could not!

At that point, the lieutenant started yelling at them. "You can't tell me that the two of you cannot pick up that rucksack. Now pick it up!" Again they tried and strained with, all their strength, but could not lift it at all. Not even a hair's width off of the ground.

Again, the lieutenant became very angry and said "Take it and put it into the truck." Then another soldier came to their aid. Now there were three grown men attempting to simultaneously pick up my rucksack, so that they could carry it to the truck. They failed; they could barely get it an inch off the ground.

At the site of this, the lieutenant became inferior rated and asked, "Whose rucksack is that? I remained silent! Again he repeated himself, "Whose rucksack is that?" Again, I remained silent!

One of the soldiers attempting to pick it up looked down to see my name tag sawn onto the rucksack and yelled out that it belongs to "Cloak!" The lieutenant then looked at me and said "Cloak, get off of the truck! Get that rucksack and put it on the side of the truck, let's go!"

Immediately, I climbed down off of the truck. Walked over to my rucksack, rolled it over on its back, with the frame pointing up! Laid down on my back, put my arms through the straps, tighten the straps down as tight as they could go. I used my M-16 as I rolled over to the right; put my M-16 butt stock into the ground! Reached out with my hands to grab a hold of the barrel and front sight, pulled myself over onto both my knees! Then up onto one knee, and then up into a squatting position. Then heaving myself up into a standing position!

Slowly, I walked over and leaned my back and rucksack/backpack up against the side of the truck. The lieutenant's mouth opened wide as his chin dropped. He could not believe what he just saw. No one in the truck, said a word, the universe was silent. And so was I! I just stood there with my back up against the truck waiting and waiting and waiting.

After a long pause of silence, the lieutenant said "Somebody get hold of the rucksack and tie it to the side of the truck! At that point 4 people in the truck grabbed a hold of it as four people got out of the truck and gathered around me on the ground pulling it from my shoulders and pushing up while the four men in the truck pulled and someone fastened the straps onto the side of the truck.

I stood there in total silence and took a long hard look at everyone's face of the entire platoon. No one spoke up, no one held their gaze to my eyes. I stared them all down, one after the other.

After a while, the lieutenant said "Okay, let's go! So I climbed up in the truck and took my position. As each of them looked in my eyes, I stared them all down, one after the other.

Later as we drove along my team leader instructed me that I would accompany Robson if we came under attack, Robson, and I was to go out the left side of the truck. Together as a team!

Note; Robson was in a different team than I was, had five years less experience then I had! And, had far less training than I had! He had just caused me to spend two days carrying a rucksack that three grown men could not lift off of the ground!

By the next morning, our truck and all other units in line with each other! Had moved up into Iraq, far enough that we were online side by side. Moving slowly toward enemy positions!

When we came within 800 meters of an enemy stronghold, they fired a mortar round at us. It landed 300 meters to our right front. Our mortar platoon behind us fired 30 at them. They all landed to the left side of the enemy position and slowly gradually inched their way over to the side of their position and then stopped.

We began moving forward, and as we came up to the enemy position our truck stopped as nine enemy soldiers came over the top of the hill within 50 meters of our truck. The command to unload was given and everyone jumped from the truck.

I jumped from the side of the truck over the rucksacks that were tied onto the side as Robson ran out the back. I immediately pointed my weapon at the nine soldiers and started yelling at them to "get down, lay down on the ground!" Out of the corner of my eye, I could see Robson running off to the north over the hill out of sight. The rest of my platoon did the same thing in all directions. I was stuck alone against 9 enemy soldiers that I had just captured.

As I stood there pointing my weapon at them, they all began to slowly lift their heads and look around to see that I was by myself. The rest of the soldiers in the truck had scattered in all directions, and they were being captured by one man, me.

As I watched them, they began to whisper to each other and get up on one knee. I could not understand their language, or read their lips as they were whispering. Even so, I understood their body language. They were getting ready to jump me.

Immediately, I started yelling and pointing my M-16 at all of them saying "No! No, lay down, no more talking," as I put my finger to my lips in a hand and arm movement to say keep quiet.

I walked to the left then I'd walk to the right. The whole time keeping my weapon pointed at them, expecting someone in my unit to come walking up at any moment.

After about 10 minutes, our platoon sergeant came walking over and said "Okay Cloak! Have them to walk past you, check them for weapons. Then point them south toward Saudi Arabia, and tell them to keep walking."

I looked to the north then to the south and could see that large groups of enemy soldiers were all walking by themselves back toward Saudi Arabia.

As I used hand and arm signals, and Stern English, I directed each enemy soldier that I had just captured to stand up one at the time. Come over to me, and empty out their pockets. Then start walking directly towards Saudi Arabia to the south.

About the fourth man that walked up to me and started emptying out his pockets, pulled out a half rotten orange and looked sadly into my eyes. I knew what he was trying to communicate to me, which was; that it was the only food he had and could he please keep it!

It was very obvious that the US Air Force had beaten them so badly by not allowing supply's to reach their location that they had been ready to give up for days.

I immediately pulled out a full unopened MRE from my cargo pocket on the side of my desert camouflage pants. I motioned with my hands that he would get one just like it, once he got to Saudi Arabia! This put a smile on his face; he threw the rotten orange down and started walking briskly toward Saudi Arabia.

Immediately, the men behind him emptied out their pockets. Walked up to me, let me check them and then ran to catch up with their buddies.

10 minutes later, Robson's foxhole buddy started calling out to me "Cloak Over Here, I Need Your Help!" As I walked up to him, he says "Cloak, we need to clear out these enemy foxhole and fighting positions. I said "Okay, let's do it! When he said, "I don't know how, you show me! I said "Yes you do, the same as we trained! He said "No, I don't know how. I missed that training! You show me!"

As I stood looking at his eyes, I knew he was lying. What he was doing, was to send me in first. That way if there were any tripwires and an explosion happened. I would be the one that got killed.

I was disgusted. I had had enough of cowards. I turned and walked away and went over to a foxhole by myself and started clearing foxholes without him.

Soon I came to a Foxhole that had its walls coated with concrete! Inside was a refrigerator with no electric hook up! There on the wall was a large picture in a wooden frame of Sodom Hussein!

I held on to the picture, as I wanted it as a war souvenir! Soon the lieutenant would see me with it, and order me to turn it over to him! Later I would find out that he had kept it as a war souvenir!

On the middle of the third day of the ground war we had worked our way to within 70 miles of Baghdad, when a sandstorm came up seemingly from the southwest.

It looked to be a 500 foot tall wall of sand that stretched as far as the eye could see in both directions. Everyone grabbed their green army ponchos and/or camouflaged army ponchos and found a place next to the truck or in the truck.

I lay down on my side to break the force of the wind from the sandstorm, and covered myself with my poncho.

When the strong winds of the sandstorm and the sand beat against the poncho it pressed against my face so hard that I had to use my hands to cup over my nose and mouth so that I would have a breathing space. The sandstorm lasted for 3 ½ hours and toward the last 10 minutes seemed to gradually over time, slow down to where it stopped.

As soon as the sandstorm was stopped the message came across the radio that the ground war was over, and to go back to Saudi Arabia. Immediately, we loaded the trucks and began getting in behind the truck in front of us.

Spotted through the desert, as far as the eye could see, beginning about 200 meters from us were 500 feet tall whirlwinds of sand. They were not going anywhere. They were just there spinning like a top, full of sand.

As we drove by, two spinning sisters about 50 meters apart were just spinning in place. Not getting closer to each other or moving away from each other.

As I thought to myself, I know what you are, you are Angels. You were sent with the power of God, the war is over, and my prayer has been answered. You ended the ground war in three days. You came and put an end to this ground war. Thank you Jesus, Thank you! Thank you Jesus, amen!

Chapter 8

RETURN

What I was referring to in that e-mail to "Congressman Tim Mahoney," at john.mac@mail.house.gov when I stated: I returned home to my hell on March 15, 1991. It was that, when I returned home from the Gulf War? No one was at the air base to greet me and welcome me home. Everyone else's family was there; they were all hugging and crying!

My father and cousin found me later at my unit area!

My "Bubby" and his mother were nowhere to be found. She had never answered my mail. I had no idea where she was or where she had been. Fear of what had befallen my little "Bubby" consumed my every thought.

The electricity to the house had been disconnected by the power company for non-payment! The food in the refrigerator and upper freezer compartment, were totally rotted, causing a horrible odor throughout the house.

After a couple of days, and while my father was outside, he seen my "Bubby" and his mother ride by in a car and pull into a neighbor's driveway. He then came into the house and informed me!

I immediately, walked out of my house, straight down the street to the neighbor's house and knocked on the door! A woman of whom I had never seen before opened the door and looked at me! I stated; "I want to see my son!" She then immediately closed the door. I then heard her walking through the home!

A moment later footsteps returned to the door, and it opened. It was my "Bubby's" mother. She stated angrily, "What do you want?" I said as I stepped inside the door, "I want to see my son!"

This was a highly tense moment, because my instinct wanted to confront her verbally and angrily. Even so, at the same time, my instincts were in conflict because of being in a stranger's home, of which I had not been invited. Noticing that she was looking, just slightly past me and into the room behind me, I turned completely around and looked to see my little "Bubby" asleep on a bed inside a bedroom.

Now, 3 years old!

I walked into the room, and as I bent over to look at my beautiful child. He opened his eyes and looked at me!

I said to him, "Bubby, do you know who I am?" He answered and said "Yes, you're my Daddy!" I then picked him up and said, "Come with me, I'm taking you home!"

I immediately walked past my wife, and carried my son down the street and back to our home. A few minutes later my "Bubby's" mother returned home.

The next day my father and cousin, left for their home's two states away!

A week later, my Bubby's mother stated, that she was going to go away for the weekend with her boyfriend. In this way, my "Bubby" and I could have time alone together.

That day at work, my unit had an award ceremony formation, of which I was told to attend. Everyone was given a "Bronze Star for Valor" during "Operation Desert Shield," but me! Then everyone was given a "Bronze Star for Valor" during "Operation Desert Storm," but me! Then everyone was given a "Combat Infantryman's Badge" for actions during combat, but me! Even though no one had ever fired their weapon or been fired at except for the one mortar round that landed far from anyone.

For me to be given a 2nd "Combat Infantryman's Badge", would mean a "Silver Star." And would go at the top middle of the "Combat Infantryman's Badge" that I earned during "Operation Just Cause!" Although I was right next to everyone in my unit during both "Operation Desert Storm and Desert Shield, I was denied any award during the ceremony, just like I had been denied after "Operation Just Cause" a year before. However, this time, I was ordered to stand in formation next to everyone while they received their awards.

At the end of the formation our company commander said, "You have my permission to highly exaggerate any war stories you tell concerning your experiences during the Gulf War! I don't even care if you make some up!"

This comment from the company commander, received a big cheer from the men in our unit, but made me want to vomit. Once again, I was being punished by not being awarded my war medals by having to stand among everyone else in my unit in formation, while they received theirs.

My two "Bronze Stars" for "Operation Desert Shield and Desert Storm" would be later sent to me through the mail, as words on a piece of paper that said, "Official US Army orders awarding "1 Bronze Star for Valor" during "Operation Desert Shield and awarding "1 Bronze Star for Valor" during "Operation Desert Storm." My 2nd "Combat Infantryman's Badge or Silver Star for Combat Infantryman's Badge" was denied even though I was standing next to and doing the same job, experiencing the same amount of danger as everyone in my unit when they earned there's.

When I arrived home, my wife had spent the day over at the neighbors' house, and I had put needed supplies into the trunk of my car. As soon as she came home, I would give her an ultimatum.

That afternoon as she prepared to leave the house for a weekend with her boy friend! I stood in the doorway of the room and looked her in the eye while stating; If you leave and spend three days with your boyfriend don't come back. Bubby and I will not be here!

She then said, matter-of-factly; Oh, yes you will, you'll be here. I can do anything I want, and you'll sit here and wait for me to get back.

As I looked into her eyes, I was stunned at how perfect her English accent was. Just eight months before when I left for the Gulf War, she spoke in broken English with a deep Panamanian accent. Much like Spangles!

I then stated again, as I called her by her first name; if you leave and spend three days with your boyfriend. Do not come back, Bubby and I will not be here.

As her eyes bore into me, she said, in a firm masculine tone, "You will be here when I get back! I can do anything I want!" I then said, "No! I will not be here, and neither will Bubby!"

As she walked out the door, my heart sank into a bottomless pit; I held my chin up with a determined resolve that I would not tolerate this.

Once outside, she got into the neighbor's car, and they drove away. Standing in the door, I watched until they were turning the corner and completely out of sight.

Then immediately, I walked into the house and picked up my "Bubby" and his car seat. I walked out to my car and put my "Bubby" inside his car seat, firmly and safely strapped into the car seat on the passenger side.

My car at that time was a solid black, Chrysler Le-Baron, 2 door Cop. 5 speed, with a turbo charged 4 cylinder engine. The windows, where tinted dark black…

Immediately starting the car, I drove down the street turning at the corner in the opposite direction that my wife had just gone and drove until I was at my father's house two states away.

One hour from my father's home. I had just turned off of the interstate highway 95. Where I had been driving for many hours at a speed of 70 to 75 miles per hour! Along will all traffic on the super highway?

I slowed down to 55, as that is what was marked on the signs. One mile further, I arrived at a sharp curve to the left, which was marked with a sign that said, “Reduce speed to 45 miles per hour”.

I reduced speed, and just as I came out of the sharp curve there was a road sign that said “30 miles per hour”… I took my foot off of the gas pedal and continued to slow down. When suddenly, the blue and red flashing lights of a sheriff’s patrol car, came on and pulled up behind me.

I was given a speeding ticket and told that I would have to appear in traffic court the following Wednesday afternoon. There, and only after I paid the fine, my drivers’ license would be returned to me.

The next morning, while visiting with my Uncle! I informed him of the traffic ticket, and that I would get into trouble once I return to Fort Bragg. If I stayed around until Wednesday afternoon! Because I was to report for duty at Fort Bragg on Monday morning!

My Uncle asked me to describe the area of the speed trap, and the officer. I did, he became very anger. Picking up the phone he dialed a number and this is what he said; “Let me speak to the sheriff! Sheriff, this is Reverend Saul’s of the First Methodist Church (In the same city). My nephew has just returned from the Gulf War. He is here with me. He has just informed me, that he received a speeding ticket from you, just last night. Now, Sheriff! I have lived in this city for 10 years and know that speed trap very well. It is used against criminals, as a lawful way of taking their driving license. And making them appear in court. To include collecting revenue for the city! Now, Sheriff! You cannot expect me to believe, that our men and women in uniform, returning home from the Gulf War. Deserve this kind of disrespect! I will not tolerate this! Yes, that's right! Yes! Ok! Yes, I will tell him! Oh yes, I will tell him right now!"

After hanging up the phone, he said to me; "The Sheriff said, for you to stop by his office tomorrow night on your way back to Fort Bragg. He will give you back your drivers' license and throw that ticket into the trash."

I wasn't all that shocked! You see, when I was around 5 or 6 years of age! I had shot at a passing car, on the dirt road in front of my grandmothers' house, with my BB Gun. The car stopped and I took to running. My Uncle later found me, he had a good "Father and Son" talk with me. He was an honest, upstanding, God fearing man!

After that weekend, my father had agreed to take care of my "Bubby" until I could get this situation at home under control. I was to report back to work the following Monday.

After picking up my drivers' license! During my drive back to our home outside Fort Bragg North Carolina, I cried, I begged my God for help.

Praying, I asked that if my "Bubby's'" mother was mentally ill, she be put in the hospital against her will, given medications that would help her. If she was demonically possessed, that the demon come out and face me, telling me what it wanted.....

Chapter 9

Talking

Returning home to an empty house, my little Bubby was at my dad's home two states away, and my wife was spending the weekend with her boyfriend. I had to report to work at my unit area inside Fort Bragg the next morning!

I did, and spent the day worrying over what was going to happen once I had gone home. Once home, nothing happened as my "Bubby's" mother did not return!

That night, I went into the kitchen cabinets. I took out a bottle of cooking oil, and a coffee cup. I poured the cooking oil into the coffee cup until it was nearly full!

Holding the cup of oil, with both hands, I bowed my head, closed my eyes, and then I silently prayed, "Dear Father God!

Please accept this oil as a true representation of the blood of Jesus Christ of Nazareth.

Please Father God, as I go around my home anointing it with oil, by the name of Jesus Christ; please accept it as the blood of "Jesus Christ!"

I then walked around inside of my home repeating the "Loads Prayer!" as if I were a broken record! Stopping at each door and each window, I would put my index finger into the oil and paint a cross at the middle top of the doorway! Then on each side and then the bottom; I did this until I had completely anointed each door and each window of my home.

I ended with a very serious, Amen!

The next day, after I returned home, my "Bubby's" mother was there waiting for me on the outside of the house and asked where he was. I reminded her of my ultimatum, she became very angry and left. She never tried to come in!

On the third day after my prayer asking God if my wife was mentally ill, that she be put in the hospital against her will and given medications that would help her. If she was demonically possessed that the demon, come out and face me, telling me what it wanted. I had returned home from work and heard a knock at the door.

Opening the door, it was my little "Bubby's" mother, my wife! I pushed the door open and stepped back, then said "Come on in!" She walked up to the edge of the doorway and stopped!

Consciously, I knew about the oil, I had anointed the doorway with and each window; she did not, no one knew but me and God!

She had stopped at the doorway and did not come in! Again I said "Come on in!" She said "No, I can't come in!" In perfect English with no Spanish accent and in a male tone of voice!

I said "Yes you can, this is your home, come on in!" She said, "No, it has changed, I cannot come in!"

A minute later I walked outside, after it was obvious that she was not coming in, and sat down on the front steps facing her. She was now standing on the ground in front of me, directly under the kitchen window of our neighbor, a few feet behind her!

As I looked her in the eye, I asked, "Vanessa, what's wrong? Just talk to me!" Right then, her bronze skinned face turned pitch black! The muscles around her eyes pulled her eye lids into slants. She replied in a deep wicked baritone male voice, "I'm not Vanessa! I'm the devil! I came from hell to drag you through hell!

At the sound of her voice, I became instantly wide awake and said, "If you're the devil, then you're a liar!"

With this, "Her head tilted back, her mouth opened very wide, her lips rolled back much like a horse when it is showing its teeth, and it laughed very wickedly;" obviously pleased with my words!

I then spoke up and said, "If you're the devil, why don't you just kill me?"

In its wickedly deep voice, it replied; "I don't have the power to kill you; only to drag you through hell! Anytime I do something with this body, you feel it!" Look at you; I have drained you like a vampire! *(Note: remember that I had lost a lot of weight from only having bread and water to eat for many months, then with all the worry I was under; I could not eat like a healthy man!)*

I was instantly insulted and said, in a commanding voice and attitude, (In Fact I Yelled!), "DEMON, IN THE NAME Of JESUS CRIST, BE SILENT!"

Immediately, my wife's month clamp tightly closed, as the muscles of her jaws strained as if trying to crush her teeth together.

Her month was firmly shut!

A slight noise came from behind the neighbor's kitchen window directly behind her about 7 feet away! As if someone had been startled and made a sudden movement!

As time passed by slowly, we just stood there looking at each other! She remained very quiet not opening her mouth and not looking at me directly in the eye!

After about 30 minutes, speaking softly, I invited her to come in the house and sit down! I tried to assure her that everything would be okay!

As we sat inside the living room quietly looking at each other, she stood up and walked into the kitchen and started looking through the cabinets. I was curious as to what she was doing, but remained silent! Soon she made her way into the bathroom. I could hear her opened the medicine cabinet, shuffling things around!

Soon she was back in the living room with a glass of water and a bottle of high-strength painkillers the doctors had given me when I had blood poisoning and was sent home on profile!

She began to pour the pills out into her hand and put them into her mouth, turn the glass of water up and swallow hard, deliberately! I immediately spoke up and said "Don't do that!"

She continued! I said again; now standing in front of her, "Don't do that! You don't want to do that!" I reached for the bottle; she jerked it away! The glass of water fell to the floor!

I walked to the phone; she followed me, we wrestled with the phone!

I knew I needed to call someone to get some help, so I spoke slowly and softly! You're going to start feeling sick, I said! "You're not going to like how it feels!"

Soon she was sitting down on the couch and turning gray!

After about 20 minutes I had her convinced into getting in the car with me and allow me to drive her to Womack Army hospital emergency room at Fort Bragg!

Once there she sat down in a chair and put her face into her hands, as I talked to the man behind the desk. I explained what had taken place!

Soon she was inside the emergency room on a table with a tube down her throat pumping out her stomach! The doctor came to me and advice that I leave her there with them. Drive down into the city of Fayetteville to the Sheriff's office.

He said, "Tell them that I sent you there to get a warrant for her arrest. So that she can be committed in the Cumberland County mental hospital against her will!"

Two weeks later, I had walked over to my neighbor's house and knocked on the door. A moment later she came and opened the door! Once noticing it was me, she immediately said; "Come in, please come in!"

I had never spoken to her before; in fact, I had never even seen her before. However, she had seen me, and soon was talking to me, a mile a minute. She went on to say that; she had been standing by her kitchen window the day my wife said; "I'm not Vanessa! I'm the devil!"

I was surprised! I was there to ask for help, watching my house when I was away! She was ready and willing to help, but wanted to make sure that I would leave her name out of anything and everything!

She was afraid of my wife!

She went on to say that, my wife's strange activity had been going on for a long time before I left for the war, and that she had disappeared not long after I had left for Saudi Arabia. Even that she had only come around on a couple of occasions shortly before the war ended.

Also, she mentioned that always during times when I was not at home. A car load of men would pull up to my home, and everyone would go inside. To include the times that she would watch through her kitchen window over through my living room window and watch my little "Bubby." Either sleeping on the couch or playing with one of the men, in the middle of the floor!

She wanted me to know that, often a car would pull up into my driveway and what was going on inside my home was not kept from my little "Bubby." To include the man in the "corvette," who would always park his car in backwards at the end of the driveway? She said; "He always parked so that he could make a quick getaway!"

If, two and two equal four; who does that sound like?

Soon, I was divorced and a single father. Winning my court case for custody of my son was not a difficult thing. As she did not fight for any rights, to see her son!

Chapter 10

Robson's Comment; Irreverent "outrageous treatment" in general!

One morning on my way to work, I had to stop on the side of the road and change a flat tire. Once I arrived at work, I was late and received a bad counseling statement from the Squad Leader, who had watched Robson and his foxhole buddy assault me back in Iraq, and did nothing about it. He even watched me get punished by my chain of command for doing the right thing, that he was part of, and did nothing to stop!

Then a few days later, we had an inspection formation. I needed a hair cut but had no money left over after paying all of the many over dew bills, which had not been paid sense I left for Saudi Arabia 10 months before. So the same Squad Leader offered to cut my hair for me. He did and my hair cut caused me to fail the inspection. Then the same Squad Leader that had just cut my hair gave me a bad counseling statement for my bad haircut.

I lost it, and gave him a bad counseling statement at the bottom of my bad counseling statement from him. He was shocked, and had no comment! Gee, this man was getting on my nerves!

To think that he gave me a bad counseling statement, for having a bad haircut, and he had been the person who gave me the bad haircut, was more than I could handle!

I was temporally insane once again!

My thought process was not motivated by fear, but my desire to respond in the correct manner to the problems before me. I would think about every little thing that happened, considering each alternative, but ultimately the answers didn't come from logical analysis. There were just too many variables! Geee, I could only think of one thing, which was; "I cannot like this man!"

Then as I walked down the hallway, I passed by Robson standing with 5 other members of our platoon. As I stopped to see what they were doing, Robson says, "Cloak, I seen your wife sucking the devils dick!"

As I looked him in the eye! I could see everyone smiling in my peripheral vision. I could tell that they had all been waiting for this chance for him to say it. Thinking that they would see a fight!

I looked at him dead in the eyes and said, *"That's nice Robson, did you enjoy yourself?"*

They were all shocked, even Robson didn't know what to think!

Then someone called out "FORMATION."

As we all walked down the stairs, I was stopped before going out the door and told to report to the company commander.

Soon I was standing at attention inside his office as he lectured me over my little Bubby. Saying that; single fathers were not allowed to serve in the 82nd Airborne, and that I should conceder giving my son up for adoption!

I was at a cross road, and needed to talk to someone! I needed to decide, as whether or not to get out of the Army. I still had 1 year left on my enlistment and would never give up my son for adoption. I could not care less what the company commander thought I should do about my child. He had no right, and no say so on this matter!

I was, temporally insane once again and knew that I did not want whatever was coming my way, for the rest of that day.

I had, had enough of their B S.

That night I was supposed to jump from a C-141 Star Lifter with a parachute packed by someone I had never met or had ever seen. So that a new lieutenant could get some training on how to lead professional paratroopers.

I went to my platoon leader, and asked to be excused from the jump. He refused my request. So I said, "Sir, I am not jumping tonight!" He replied, "Yes you are, report to pre-jump at 17:30 hrs.

I did, and walked up to him and said "Sir, I am not jumping tonight!" He replied, "Yes you are Cloak, now get into formation!" I did, and after pre-jump practice, I walked to my car and drove home! He should have known that I was a man of my word; I had never lied to him before!

That night I drove to Wilmington North Carolina, and walked into the emergency room of a hospital. There I asked to talk to a counselor! After talking to the counselor, she said that she would have to call my unit; as it was required by law. She did, returning to me to say that my unit commander had said; For me to report to my unit area the next morning…

I was temporally insane even again!

My thought process was not motivated by fear, but my desire to respond in the correct manner to the problems before me. I would think about every little thing that happened, considering each alternative, but ultimately the answers didn't come from logical analysis. There were just too many variables! I could only think of one thing, which was; "I will be taken directly to Leavenworth prison!"

The next morning, once I had reached my unit area, I was summoned to the office of the company commander. Once standing at the position of attention in front of his desk, he simply said to me, "Cloak, don't do that again!"

I knew I would, if ever I needed to, but replied to him with "Yes Sir!"

Chapter 11

Hospital Nurse; A Camel That Broke the Straws Back!

I needed to talk, and went to the medic's. They directed me to an office down the street. There I told a concise portrayal of the difficulties I was having and was told that I may go to a "Ward 4" at the hospital. There I could get plenty of sleep over the following few days and that after a few pleasurable nights of sleep; I could best think through my problems.

Once there, I was stripped of uniform and all identification, told that the door was locked for my own protection, and if I needed anything to just ask the nurse!

I still had not gotten back the weight that I had lost over in Saudi Arabia a year earlier. I was simply still fighting a war! The war was inside of me! You can visualize me as thin as a straw!

During my first day there, I was interviewed by a doctor. I told him about my many problems. At the end of telling him about my wife and the day she was the devil! I asked if he believed in demonic possession!

He answered, "I have not ruled it out!"

I slept through the rest of the day and most of the night!

The next morning was uneventful, as I was walked down to the hospital commissary for breakfast by a male orderly of the hospital wing! I then walked back to "Ward 4" and was told to just sit around.

I was a smoker and as you might imagine, in "Ward 4" there were no smoking areas and because the door was locked, no patient's were allowed to leave the "Ward," without an orderly accompanying them. Each time I asked, I was refused, and told that the orderly could not accompany just one person!

I complained that I was the only smoker of the 3 patient's to no avail!

By the middle of the next day, I wanted a smoke, and I wanted it right now!

After complaining to the nurse, I realized that I had been lied to and was in fact under arrest, being held against my will! At this, I complained intently to no avail! Being told that I was to attend a group therapy session.

The next day, 3 more men were in the ward when I was walked down to breakfast. Each seemed to be care free of any problems but dressed in the same "Ward 4" patients' attire!

Around 9:30 AM, I was sitting among 7 men and one counselor! He started out the group session, by having one man that I had never seen before start out telling a short version of his problem.

He said that, "His wife had cheated on him with another man. It hurt him to know that this other man had been on top of his wife in his own home and in his own bed while he was at work." Then he said, "What hurt even more, was that she liked it, and all the men he worked with knew about it and never told him!"

The counselor then turned to a different man and asked him what advice he had. The man replied, "I don't know, then turned to look at me!" The counselor then turned to me and said, "Look at him and tell him what you think!"

I did and said very derogative, "I don't know why you're complaining, your wife only slept with one man, big deal! If that was all I had to worry about, I wouldn't be wining like a little baby! My wife slept with many men all at the same time, and all of them worked with me or over me in my own unit. She says she did this simply because she wants me to feel pain. Hell my own First Sergeant had me pulling 24 hour duty so that he could go by my home and stand in line. If I were you, I'd just shut up, go home and live with the knowledge that your wife only slept with one other man, not so many, all at the same time!

And to top it off, my 3 year old son has to go through the rest of his life without knowing his own mother because she is unfit to be around children. The government child protection agent assigned to my case, told me that if I ever let my son's mother see him again, she would take my son away from me, and she would never let me see him again! I would never know if he was dead or alive! So maybe you should just shut up!"

At the sound of this, all men in the room were totally silent! The counselor didn't even know what to say!

The group session ended and the men in that room were never seen by me again. They just disappeared!

By the end of the 4th day, I wanted a smoke. So I called my unit area, and asked for the man that had been overburdened back during the Gulf War, just like I had been. By being made to carry everyone's equipment for them into Iraq! He said that, he would try to get by there to see me that night!

The phone was way down at the back end of the hall way. As I was returning to the TV and sitting area, the phone rang. I did not think much about it and continued on walking back to sit in front of the TV, thinking that someone would answer it.

After it had rung for 5 minutes, and no one acted like they cared! I walked back to it and answered it, "Hello!" a woman said, "Is Mark there?" So I called out, "Mark, Mark, you're wanted on the phone!" When no one came, I said to her, "There is no Mark here; you must have the wrong number!"

She then asked what the number was at this phone, so I read it off to her. She acted surprised and asked where it was that I was answering the phone from. So I told her the truth, which is, it was "Ward 4" of the army hospital at Fort Bragg.

She began asking questions and because I felt like talking, I answered her. It soon became evident, that she was CID (Central Intelligence Division) using a pretext, so that I might open up and talk to her! This also made me suspect that all the men in group session were ether CID or plants by the doctor's, as a way to gather info, as they were not patients and never came around again during the 2 weeks I was there! During the rest of those 2 weeks, she called everyday!

Even though I was being treated as if I was insane, my thought process was not motivated by fear, but my desire to respond in the correct manner to the problems before me. I would think about every little thing that happened, considering each alternative!

Soon my friend came by for a visit and brought a pack of smokes. Passing them to me covertly along with a Bic lighter!

Each morning, when I took a shower, the smokes went with me. I turned the water up as hot as I could take it, so that the room filled with steam. This allowed me to smoke covertly over many days and nights as I began taking a nightly shower just so I could smoke.

There were never anymore group therapy sessions or visits by a doctor. Each morning and each afternoon, the phone would ring, and I would answer it. Always the same woman, who now seemed to think that we were friends! The more in-depth her questions the more I was sure that she was an investigator using a pretext. So I made a firm decision, that I would let her in on my smoking in the shower, which was against "Ward 4" rules. To just see, what transpired!

The next morning, 30 minutes after my shower, I was taken into my room and given a strip search by a Staff Sergeant E-6.

As I stood there naked, I said "What are you looking at? Are you gay?"

He said, "You have been smoking, give them to me!"

I did and he left the room! 3 hours later, I was told to go into the 3rd office door down the hall and talk to the nurse there.

Once I walked into the room and looked at the face of the nurse, I knew that I had seen her before and asked, "Do I know you?"

She immediately stated; "Do you think that my daughter and I should have to go through our life, knowing that there is something like you, walking around!"

I was shocked, "I knew I had seen her before!"

She went on to say, "My daughter, and I have to live in this world, we have to go shopping at the mall. Do you think that we should have to put up with something like you, in the same mall at the same time we are there?"

Angrily, I looked her in the eyes and said, "During 1985, I was in this same hospital for lock jaw. Early one morning I opened my eyes, to see you doing something with my finger while you thought I was asleep!

"She looked like she had put on a few pounds sense then!"

You can imagine her; the size of a camel!

Recognizing me, she became teary eyed and said "When I get through with you, you will never get out of here!"

She then left the room!

I was disgusted, and went to the front desk! There I told the nurse to call the doctor and tell him, that "I was refusing any more treatment from him or this hospital and that I wanted the door unlocked, right this moment!"

She said, wide eyed and shocked; "OK!"

I then waited 10 minutes, and as she got off of the phone, I walked up to her and stated, questioningly, "WELL!"

She replied, "The doctor will be by to see you at 2:00 PM!"

I did not go to lunch, but waited, for the moment, when that doctor would walk through the door.

At 2:10 PM he walks into the "Ward!" I was told to wait until he called for me. Once he sent for me, I walked into his office and began to tell him what the obese nurse had said and done. He stopped me after the first sentence, and said derogative, "I don't care!"

I then spoke up loudly saying, "I am refusing, anymore treatment from you and this hospital! You will unlock that door, and I will walk out of here!"

He then asked, "What do you want?"

I said again, "I am refusing any more treatment from you and this hospital. You will unlock that door, and I will walk out of here!"

He said, "No, just tell me, what you want?"

I replied, "I just told you!"

"No," he replied, "I mean, what do you want when you get out of this ward?"

I told him sternly, "Out of the Army!"

He repeated me as if surprised, "Out of the Army?"

I replied firmly; "YES!"

"Ok," he said, "I can help you do that!"

After two moments of silence, I said, "OK, you do what you can to help, I want out of the Army! I have a young son to rise! I also want out of this hospital today!"

"OK," he replied, "I will need a couple of hours, then the door will be unlocked, and you can go back to your unit area!"

The door was unlocked at 4:30 PM! I went back to my unit, and waited!

Well, they had done it! They had destroyed the career of this Special Forces Trained Soldier. Still, it took the Devil and the entire 82nd Airborne of the US Army, many years to do it!

And, of course; A Camel That Broke the Straws Back!

Life can seem to be cruel, can't it?

30 days later, I was headed north on highway 95 with a fully "Honorable Discharge", to get my little Bubby! I have never left him, and am still with him even today!

You now know what I was talking about in my email to the office of "Congressman Tim Mahoney" at john.mac@mail.house.gov about;

1.) The officer appointed over me during "Operation Just Cause," who gave out all of his equipment to soldiers under him to be carried by them into combat, so he could fill up his backpack with a pillow.

2.) Dreams that come true.

3.) Chop to the throat by a fellow soldier.

4.) Met and talked to the devil.

Those of you that were married to members of the 82nd Airborne during that time in history might wonder if your husband really was at the place where he told you he was, and rightfully so, for "Rank Has Its Privileges!"

Further, I understand that it is throughout the entire US Military, even today!

Too bad, I had not gotten in touch with "Joe Alercia" and his "Lion Investigation Academy" way back then, isn't it!

So read on; now that you have gained a more in-depth perspective; and give a good thought to a class action law suit to everyone involved; so that they have to repay all tax payers of the United States of American! The many hundreds of thousands of dollars that have had to be spent on my 100% Service Connected PTSD!

Not to mention the many hundreds of thousands of dollars that were spent to train me, just so they could hate me for it and work covertly to destroy my future!

With their selfish, self centered cowardly betrayals!

Chapter 12

DANGER

"When I looked at his face, into his eyes, it was instant recognition, and he knew it! I had seen him before!" It was "Osama bin Laden". Take note of the color's they were wearing! They use color codes; i.e. "Orange and Blue!" During the history of the "War on Terror," who was wearing orange?
Later in this book you will read the exact first message, sent to President Bush through email, concerning my encounter with "Osama bin Laden"!

On December 2, 2007, my "Hun" and I had rented a car and drove to the SM Mall in Pampanga on the island of Luzon here in the Philippines.

On the way there, I was thinking to myself that I should offer to buy my wife a new cell phone as her old one was giving her many problems.

Once we had parked and gone inside the mall, I stated to her "Let's buy you a new phone!" She liked the idea, and as we walked through the mall looking for a cell phone of her liking, we came upon the public restroom.

She went inside as I waited outside! With as many people going in and out as there was, I needed to step over to the side out of the way. Doing so, I put my back to the wall....

As I stood there, I watched the security guard over at the mall entrance as he checked people's bags when they came in the door. A female security guard would check the women's pocketbooks.

Many people were going in all directions. Suddenly, a man in a white buttoned up shirt, tailored black pants and a very short hair cut, came walking up the far side away from me into the open space between people walking.

He looked directly into my eyes and stood there for a moment. Instantly, I knew that he was there to watch me and was most likely, mall security. So I looked for a radio or other communication device in his hand, down by his side.

There was not one in his hand or on his belt!
Right away, he glanced up on the wall over my head, which caused me to think that he would be writing a report and needed the time as I had already glanced up there at the clock myself!

This confirmed in my mind that he was working in mall security and was keeping me under surveillance.

I did not feel this was a problem as I was not up to any harm!

At once, he turned back in the direction he came from, and started walking away down the mall. Which was to my left, the wall was to my back, and the public restroom's entrance was on my right side?

As he walked beside a group of people walking together, I lost sight of him. This put a question in my mind immediately, because I was looking forward to seeing him come out on the other side of them. So that I could follow his activity!

He did not come out from behind them and there was no door within 20 feet of where he disappeared behind the group of people.

Because there were only five in that group, I felt that I should be able to have caught sight of him as he walked past them, easily!

I took notice of the next door in the direction he was walking, and that it was a business selling cell phones. The thought was in my mind that he had very possibly walked into that business!

A few minutes later my wife was standing beside me, so I said "Let's look over there and see if they have a phone that you want!"; while I was pointing to that same store, I believe the man had gone into. I was thinking that I would see the man in there and that would explain where he had disappeared to!

Once inside the store my wife started looking at different phones. I looked around expecting to see the man in white shirt. I did not, but there was a door behind the counter, which he could very possibly have gone into. Which is what I was thinking?

The man behind the counter was younger and thinner than the man, I had seen in the white shirt, also he had longer hair.

He had asked my wife, how he could help her. She was asking him, for information about a particular cell phone!

He stated that the phone costs 4,500 pesos!

As my wife looked at me to see if I was agreeable, I pulled out my wallet and started counting the 4,500 pesos, in 500 peso bills.

I had 8,000 pesos in my wallet all at 500 peso bills, and as I counted my wife watched intently. 500, 1000, 1500, 2000, 2500, 3000, 3500, 4000, 4500!

Then I handed the 4,500 pesos to my wife as the man behind the counter retrieved a box with the new cell phone my wife was now buying. As she stood at the counter across from the sales clerk, I stood next to her. She began counting the pesos again as I kept up with her, the whole time counting along with her in my mind! 500, 1000, 1500, 2000 and on until she had also counted 4,500 pesos!

The money never left my site!

Once the clerk had finished testing the phone insuring that it worked properly and allowing my wife to also check it to see that it was working properly, my wife handed him the 4,500 pesos. Again the money never left my site!

The clerk began counting out the money on top of the counter in front of our eyes. Again, it had never left our eyesight. He counted up to 4,000 pesos!

I was shocked, as this was unexpected!

My wife also was shocked, and spoke up saying "No, it is 4,500 pesos!"

The clerk counted it again, right there in front of us! Again he counted 4,000 pesos as we counted right along with him!

Immediately, my wife took a hold of the money and counted it out loud. She also came up with 4,000 pesos. At which time I spoke up and said "No, I counted out 4,500 pesos!"

My wife also spoke up and said "I counted along with you as you counted 4,500 pesos!" and "When you handed it to me, I counted 4,500 pesos!"

I immediately pulled out my wallet, knowing that I had 8,000 pesos in it before I counted out 4,500 pesos, and sure enough I only had 3,500 pesos left.

I looked on the floor to see if maybe some how it had fallen. The whole time in disbelief as the money had never left my site.

My wife and I, are very good at counting money!

So I spoke up and said "I counted 4,500 pesos! Then I counted as you counted 4,500 pesos!" Then the man behind the counter spoke up and said "Yes, I counted along with you, but there are only 4,000 pesos here!"

I spoke up in disbelief and said "Well, where did the other 500 go?" The whole time, my conscious mind was in disbelief because I never saw that man move, his hand from the top of the counter, while he was counting the money! It was an empty counter, meaning there was nothing there for him to slip the money under or behind! I never saw my wife's hands leave the money as she held it in both hands, on top of the counter!

Again, my wife counted it slowly, one 500 peso bill at the time making sure that no bills were stuck together and came up with a total of 4,000 pesos!

I then took the money and counted it and only came up with 4,000 pesos! I then handed the money back to my wife and pulled my wallet out again to again find only 3,500 pesos!

All three of us looked at each other in disbelief!

Not wanting to make this into a scene and have security called on us, I immediately pulled out another 500 pesos and give it to my wife! She then paid for the cell phone!

Once we walked outside of that store we stopped and started talking about the missing pesos. I stated that I kept my eyes trained on that money from the time it left my wallet, and I counted 4,500 pesos.

My wife stated that she counted along with me when I counted 4,500 pesos. Then I stated that when she had it in her hands and counted the 4,500 pesos; I also counted along with her and counted the 4,500 pesos!

We were in shock and disbelief, neither of us believing what had just taken place.

I stated, "A glitch in the matrix!"

Two and two equal four; something is going on with my money!

On Dec 5, 2007, "Hun" and I had rented a car and drove to the "Bureau of Immigration" offices in Manila Philippines. I had to go down the street to a clinic and get a blood test and chest x-rays, for my permanent resident visa!

We had to return the next day on the 6th for the results. Then, pay towards the 37,000 pesos we were paying partly today to the man from immigration.

I then went to the US Embassy to confront the Social Security Administration case worker who was dragging her feet with my case! She was not there, so I waited until my number was called. Hoping that she would show up! Still, when my number was called, I had not seen her, so as I walked up to the window, I yelled, "This is for [redacted]

I wish that I had thought to do that much earlier, as they came alive. Reaching out hands from all directions; saying all together, "I'll take that!"

Funny thing though, [redacted] does not work there anymore!

We were to return to the Office of Immigration on the 10th of Dec 2007 for the last part of the paper work!

We arrived at the Bureau of Immigration early, and walked across the street to a restaurant for breakfast.

If you have Google Earth on your computer, go to Manila, Philippines and zoom in on: 14 degrees 35'35.43" N, 120 degrees 58'28.07" E / there you will find the Office of Immigration. I believe it was some time during Oct or Nov of 2007. When the Office of Immigration moved from its old location a 14 degrees 35'39.69" N, 120 degrees 58'31.43" E / the clinic where my blood test and chest x-rays were taken, is at 14 degrees 35'34.87" N, 120 degrees 58'23.99" E / the restaurant where we ate our breakfast is at 14 degrees 35'37.64" N / 120 degrees 58'26.64" E.

After finishing breakfast, "Hun" and I walked over to the Bureau of Immigration and just hung around. After about an hour, my Hun sent a text to the man from immigration on her cell phone! After a few minutes, his text back said that he was waiting for the results of my blood tests, from a few days before. A few minutes later he sent a text to us, saying that he needed to meet with us so that we could go have my fingerprints taken!

30 minutes later he showed up with another man we had never seen before, and asked that this man ride along with us so that this man could give us directions while he stayed there and finished his work. I assumed this was because he needed to be in two places at the same time, and the extra money, he was charging us for my permanent visa could easily pay for this man to ride along and give us directions.

Which was an arguable point for him charging the extra 2000 pesos? I doubt very seriously that the man who road along with us to give us directions received more than 150 pesos.

We proceeded north and a couple of blocks later pulled into a parking lot. Once we had paid the parking lot attendant, we walked out to the street and turned on the sidewalk.

Following the man at a brisk pace, we crossed through traffic. Walked down sidewalks, and crossed through a store. Then walked out the back door of the store onto another street, crossed that street. And, as you can imagine, I was pretty much turned around. I had no idea how or where I was.

If you have Google Earth set up on your computer! You can find the location in which we parked the car, and followed the man to the building in which I was to be finger printed.

Open Google Earth, Go to Manila Philippines.

Put your mouse over 14 degrees 35'38.28" N and 120 degrees 58'31.32" E…

Then zoom in, and you are there….

Suddenly, we were in a building walking through a door. We walked up one flight of stairs, and then up to a second flight of stairs. Walked out into a lobby and turned right. There, at the far end, was a counter with a man in a police uniform, sitting behind it.

As we walked up to the police officer, the man who rode along with us and led us to this building said, "Just sit right there! Pointing to a group of chairs! Then walked up to the police officer and started talking in the Philippine language of Tagalog.

My PTSD was highly active along with my high blood pressure.

I did not feel comfortable with the situation!

The police officer looked at me with disgust!

After a few minutes, the man that had led us there came over and took me with him. Following him, we walked over to a door and stepped inside. Again he spoke in the Philippine language of Tagalog, and was discussing something with the plainclothes police officer!

In this office was an old wooden desk, with an old wooden chair. The walls had not been painted for over 20 years. Old and dusty, not what you would expect for a police department!

After a few moments, this plainclothes police officer stepped from around the desk and said in English to come with him. Following him, we stepped out of his office, walking toward another office and another old wooden desk. From within a drawer in this desk, he pulled out an ink pad and a piece of paper that had predetermined spaces for each fingerprint.

After filling out the form and signing my name. The man roughly took each of my fingers, pressed them into the pad and rolled them in the correct space on the form. As soon as he had started, he was finished and stated; Okay that's it, you can go!

The man that had led us there was now leading us back to the parking area, and our rental car. Once at the car, we drove back to the office of immigration, and text the man from immigration that had been assisting us. At that time, the man that had led us so that I could be finger printed, said goodbye and left the area.

After texting the man from immigration we received a text back that said we could go home. However, that we needed to come back tomorrow after the police department had run a check on my fingerprints!

-- -- --

Early the next morning, we were late waking up. Combined with all the pedestrian traffic and tricycle traffic we arrived without a moment to spare. There we met with the man from immigration and followed him through traffic and down towards the US Embassy, which was south of the Philippines immigration office.

Suddenly, we were making a right turn that headed out towards the ocean. Then just as suddenly, we were making a right turn into a gated parking lot. There a guard opened the gate and pointed where we should park. Once I parked, we followed the man from immigration and walked through an office door. Once inside, the man from immigration spoke in the Philippine language of Tagalog and explained why he was there and why we were with him.

A few moments later he turned toward us and said we could go. Once outside in the parking lot, we asked him what was going on. He said that everything was okay; it is just that their office is going through renovation. Because of the renovation, all of their files and equipment are in storage. This means that we will have to go through this process at a different office, and that they had given him the address of the new location. He went on to say that we would not have to go with him; he would be able to do it on his own.

It was now, close to midday. We left and went to a restaurant to eat. The restaurant was not far from the office of immigration. So we had walked there after parking the car back at the office of immigration.

This restaurant was located at Google Earth: 14 degrees 35'36.62" N and 120 degrees 58'26.05" E.

After sitting in the restaurant for well over an hour and a half! The man from immigration text My Hun and said that all the paperwork was ready and for us to meet him at two o'clock.

He did not show up!

At three o'clock we met up with him on the outside of the immigration office. He was carrying a folder filled with forms and said; "Come on, follow me! I know a place we can do this and not be disturbed. You don't mind sitting outside, do you?"

As the man from immigration led the way, we stepped in behind him and casually walked straight across the street from the office of immigration up onto the sidewalk.

We continued forward and passed by the front door of a bank. A short distance past the bank we turned right into an area that had picnic tables with large umbrellas covering them.

It was the outside of a coffee shop! I do not remember the name of it, but right next door was a Star Bucks!

As we turned right into this area and walked along the side of the tables, I stepped over to the right side and tried to keep up with the man from immigration, as he was walking with a purpose.

Directly, in front of me, about 20 feet were two women and two men sitting behind a table, with their backs to the outside wall of the coffee shop! The table was in front of them as they looked over it. Seemingly, as if they were waiting for someone and wanted them to see them or find them easily! As if, they wanted to keep an eye out for someone that they may have been waiting for!

I looked straight at them, the younger of the two women was on the far left side looking straight at me, to her left was the older of the two women, looking straight at me, to her left was the older of the two men, looking straight at me, to his left was the younger of the two men, looking straight at me. All as if they were waiting for someone they knew to walk around the corner and see them!

They would not have stood out as so obvious to me, because they were dressed just like any American on vacation. Except for the fact that I had seen one of them before, long ago!

The two men had short hair cuts, totally clean-shaven faces. The younger of the two men had on a solid bright **orange** baseball style cap. The same type of clothing I had seen all of my life in America and in Panama. Here in the

Philippines also, it's just common everywhere to see Western-style clothing.

Note: Take notice of the colors! And who wears what! Our people who were beheaded, had to wear **orange**, right? And the terror prisoners at "Guantanamo Bay" did too! Now that I am writing this book, after the many months that took place sense my "SALUTE REPORT" concerning my encounter with "Osama bin Laden" and the changes in the color of terror prisoners at "Guantanamo Bay" uniforms. White and blue is the colors that al-Qaeda uses now, when they want to be found by their peers!

That day, what stood out about their clothing was that it was obviously brand new, straight from the showroom floor. Especially the women's, only the younger of the two men was wearing a pair of blue jean style pants that was just a little away from showroom floor quality. They would fit in anywhere, where tourists from a Western Civilization were hanging out!

Look around you, would you know them if they were in sight? You might have been sitting across from them at a restaurant!

As we took a few steps closer, my eyes met the eyes of the older of the two men. Instantly, my neck straightened, my eyes opened wider, my inhaling of breath paused. Without giving it a second thought, I knew him!

"When I looked at his face, into his eyes, it was instant recognition, and he knew it! I had seen him before." It was "Osama bin Laden".

Instantly, his neck straightened, his eyes opened wider, his inhaling of breath paused. Without giving it a second thought, he knew I recognized him!

I was tired from my full stomach, and the Philippines hot climate. However, I became instantly energized by this situation. I was wide awake, and in total control of my senses.

My senses were not motivated by fear, but by my desire to respond in the correct manner without getting killed! I was thinking about every little thing that happened, as it happened, considering each alternative, and then moving into the next instant, as natural as possible, bringing all my many years of experience to bear!

Even the hair on the back of my neck instantly became radar. Just like it had morphed into many years before, after a dangerous encounter that I didn't see coming!

I was a young man at the time, maybe 20 or so, working as a backhoe operator in an orange grove, in south Florida. I had completed my entire tasks with the backhoe and had been temporarily assigned to drive a John Deere tractor that pulled a herbicide machine.

Around 12 o'clock, lunchtime, I and three other tractor drivers stopped for lunch. One of which was pulling a herbicide machine just like I was. The other two were pulling large bat wing mowers. They could mow the full distance in between, orange trees, while making only one pass down the row in-between them.

After pulling all our tractors up near each other. We grabbed our lunch boxes and gathered in a shady spot out of the sun. As we all sat around eating our lunch, we shared jokes and were enjoying each other's company.

During the months before, from time to time; I and the other young men in my age group would get together after eating our lunch, and spend 5 to 10 minutes wrestling, before we went back to work.

This particular day, it was a young man, around the same age as me, which I didn't know very well, that was operating the other herbicide machine!

After we had eaten our lunch, we sat around talking. He put down his coke, stood up, then walked over to me and said; "Let's tussle!"

I looked up at him as he stood there in front of me and repeated him asking, "You want to tussle?"

He said "Yes, standup let's tussle!

Standing up I looked at him in the eye and said, "This is just for fun right, you're not mad at me or anything, are you?" He went on to say "Were just testing each other, come on let's just tussle; I want to tussle with you!"

As we walked out away from the shady area and into the middle of the row of orange trees! We squared off; immediately, he came at me and put both his hands on both my shoulders. I also put both my hands on both of his shoulders.

When we began to push and pull trying to overpower each other, I knew what to do! Experience had taught me that once I am being overpowered and pushed to the ground, if I'll just twist really hard and pull. I can reverse the situation, putting the person pushing me down on the bottom and me ending up on the top!

I did this and ended up on top of him, with both my hands pressing down with my full body weight on both of his shoulders. As he squirmed and struggled to escape, I shifted my weight to the side of his body that was lifting.

This way, I could counter any movement he made by basically overpower him with just my body weight.

After a couple of minutes of him struggling, he stopped and looked up at me while saying; "Okay! Okay, it's over, get off of me!"

I asked, "Are you sure? He answered "Yes, I'm sure, get off of me!" I then said; "Okay, I win you lose, right?"

He said "Get off of me!" I said again, "You lose, I win, right!"

He said "Yes yes yes, you win! Now get off of me!"

Getting up I reached out my hand to grab his hand and pulled him up as we both looked at each other for a moment, and then he turned to walk away.

Okay no problem, it was over. He seemed to be a little tense about it, but as many times as my friends and I, that work there, had been wrestling during lunch hour and no one had lost their tempers. I assumed automatically that this was the end of it and went back to sit down by my lunchbox, in the shade. It was just within 10 feet from the other two men who had been watching us wrestle.

A couple of minutes went by, and then I heard something behind me. Immediately, I felt something grab at the hair on top of my head and jerk it back, at the same time a cold sharp piece of metal touched the front of my throat just inches below my chin.

I was shocked, and because I was sitting Indian style on the ground with both my feet crossing each other! I was not able to resist him!

His body weight pushed against the upper part of my back as he jerked my head backwards with his left hand by grabbing my hair on the top of my head and pressed the knife against my throat with his right hand.

I could see through watery eyes! That the two men who were sitting a few feet away, in front of me. Were both wide-eyed and frozen, while looking in my direction.

This young man, who had the knife at my throat said; "If you ever outdo me again, I will kill you!"

I said "Okay!"

He said, "I mean it, if you ever out-do me again, I will kill you!"

I said with tears in my eyes. "Okay!"

I could see through my teary eyes, the two older men watching, were still frozen in shock. And, that they did not dare move, in fear that he might cut my throat. Or, that I might try to struggle and get away, which could cause the knife to cut into the flesh of my throat.

Again, he stated firmly, "Don't you ever forget what I am telling you! If you ever out do me again, I will kill you! Do you understand?"

I said "Yes, I understand! I will never outdo you again!" He then said; "I mean it!"

I said; "Okay!"

This immediately ended the same as it started. He let go of the hair on my head, moved the knife away from my throat, took his body weight off of my back, stood up and walked away.

Not moving, I continued to look through teary eyes at the two men who watched this entire situation from beginning to end. As they followed his steps on the other side of the orange tree as he walked away and got on his tractor, cranked it up and drove away.

This young man's character was now painfully obvious!

I sat there in shock! Not sure what to do. Even so, as his tractor sounds had become further and further away. As the sound of it became less and less. I became angrier and angrier, then stood up.

I grabbed my lunchbox; put my things back into it. Walked over to my tractor, climbed onto it, put it in gear and turned it in the opposite direction he had gone. Driving along, I changed into a higher gear, and then into another higher gear; until I was in its highest gear and had the RPMs at full throttle.

I drove nonstop as fast as that tractor would go, the 2 miles through the orange grove back to the workshop and onto the parking area where I had parked my truck.

Pulling up beside my truck, I stopped the tractor and put the gears into neutral while jumping off of the tractor. Walked over to my truck, unlocked it opening the door. Reached behind the seat and grabbed my 22 Magnum caliber pistol that had a 9 and 1/2 inch barrel.

I was going to confront this young man, and I was going to do so by pointing a loaded 22 Magnum pistol straight at his head. I was going to tell him that he had better leave, "You better leave now!" and he had best never returned. Because, if he did return, I would kill him! And, I was going to kill him now, if he did not immediately leave, never to return.

Before I could get halfway back to the area where the incident happened, our foreman came squealing up pulling in front of me with his company pickup truck. Immediately, I had to slam on the brakes to keep from running into the truck.

He got out of the truck and walked around to me and said "Get down, on down here with me!"

Climbing down off the tractor, I looked up at him as he was taller than me.

He said, "Give me the gun!" I did not say anything, my teeth and mouth were clamped shut.

He moved his hand closer to me and said "I said to give me the gun! I know you have it! I heard what happened! I've already found him and sent him away from here! You will never have to worry about him ever again! Now give me the gun!"

Looking into his eyes, I knew that he was telling me the truth and that the situation was not going to go any further, actually I was relieved.

*****

You now have an appropriate perspective, and you are with me!

They had been hiding in the open all along!

As "Osama bin Laden" and the younger man immediately stood and stepped away from their chairs. They came around the table toward us. The whole time looking directly at my face with deep heartfelt concern written across their faces... As if this was to be a confrontation, and they were going to meet it head on by standing and moving out in the open away from the table.

As they came around the table and in front of it to meet us, I eased my eyes to the left, knowing "My Hun" was with me and in danger. While keeping them in my peripheral vision, and acted as if they were not there, or as if they meant nothing to me.

As the man from immigration led the way, we walked up, straight towards them and then stopped, just to the left, and to the front, one table from them.

The man from immigration immediately picked a chair with his back directly to them. I wanted to be facing them. I, in no way wanted them behind me out of my site! So I walked around to the left of the table, while my wife walked around past the man from immigration to the right and sat next to him. She was now sitting and looking back the way we had just come from.

Continuing to the left, I walked all the way around to My Hun! The chair on the right side of my wife had a thick layer of clear water in the seat. I grabbed hold of the back of the chair and lifted it, leaning it forward and poured the water out onto the concrete. Then set it back down easily!

At the same time, keeping the two women at the opposite table in my peripheral vision and "Osama bin Laden" with the younger man, in my eye's outer peripheral vision behind the man from immigration across from me!

I instantly knew that the wet seat had attracted everyone's attention, and that I would have to go further and find something to wipe it dry, before I could sit down without bringing further attention to myself.

I could feel the people behind me to the left, inside the coffee café looking out at us through the glass windows.

To sit down in that wet chair was wrong; I could feel in my soul that if I sat there, everything would go very wrong! My memory was at peak performance, as the words "Act like; you do not know what is going on!" were clearly at the front of my thoughts.

I immediately turned back to the right and walked around the table and sat down in a chair to the left side of the man from immigration. Knowing this fit in perfectly as a pretext, I could relax and go with it, even though it put my back to the now standing "Osama bin Laden" and his bodyguard!

I knew it was right, because it fits in perfect with "Act like; you do not know what is going on!" The man from immigration and I would be speaking directly to each other while I filled out the forms, and he explained each and everyone. Also, we would be giving him the remainder of the 37,000 pesos which we had given him part of the day before, fit in perfectly as a pretext!

This worked, as "Osama bin Laden" did not come any closer to me. In fact, he and the younger man had not moved any further than three or four steps away from the table they were originally at, which left the bodyguard standing about 3 feet directly behind me!

As the man from immigration began to pull out forms and explain them to me. I haphazardly listened to him just enough to be able to nod my head in answer of any statements he might have. Which would require my communicating that I understood?

While he pointed to the form, I casually moved my head to the right and looked directly at the younger of the two women who had been at the table with "Osama bin Laden".

She was a middle aged woman, who had kept a youthful appearance. At first glance, it was, as if she were in her mid to late 20's.

Or!

At first glance, she was a beautiful middle aged woman!

With a closer look, it showed though that she was very well to do. As if she had never had to work for her entire life, or as if she had been in a health spa 24 hours a day, 365 days a year for her entire life.

She was stunningly beautiful, with the same family nose and bone structure as "Osama bin Laden". The skin on her face and her arms was flawless. Her hair hung down past her shoulders and was as lusciously silky as anyone's hair I have ever seen. It was long and flowing, shining brown in the midday sun. It was as if I was looking at a top model in a fashion magazine.

Again, I must emphasize, "She was stunningly beautiful!" The dress she was wearing was a fine silk looking dress with short sleeves and a high cut, rounded and seamless collar. The dress was multicolored and of very high quality, surely silk. It reached down almost to her ankles.

In fact, she captivated me as I barely have a memory of what the older woman next to her looked like. I do remember that she had the same family nose but was older and wore her hair under a head scarf. Although I have limited visual memory of the color of her clothing, the younger woman dimmed her out.

While the man from immigration continued to turn pages of the form, I looked at the woman who was so stunningly captivating. I gradually turn my head to the right to where I could see "Osama bin Laden" standing behind me and the man with him staring at the back of my head, as if he were ready to jump me.

It made no sense, he had me easily, if he wanted to attack, and I knew it. Although I felt no nervousness at all, I still thought to myself, "This makes no sense; they have to know that I'm acting like there's nothing going on!"

I casually moved my head, back in the direction of the sound of the man from immigration's voice. And acted as though I was unconcerned about what was happening behind me.

In my peripheral vision; I could see my wife with her head lowering. And I knew, she could see that these two men behind me were acting inappropriate to anyone that did not recognize them.

This may be what curved the situation into my favor, because no one recognized "Osama bin Laden", but me! To them, he was just a well-dressed Islamic man that obviously did not like me, an American.

As I continued to listen to the man from immigration, I kept the two women in my peripheral vision. I could tell that they were concentrating on "Osama bin Laden" and the younger man who was behind me. Again, I moved my head to the right and looked through my peripheral vision over my right shoulder and noticed that "Osama bin Laden" was pacing between the younger man, who was obviously a bodyguard, and the table where the women were. Just then he pulled out of his pocket a cell phone and walked out past the bodyguard and toward the street out of my peripheral vision.

As I turned my head to the left and continued to listen to the man from immigration, I turned my head as far left as it could go and looked to the far left near the street and the parked cars by the sidewalk. Again "Osama bin Laden" lifted the phone up, dialed a few numbers and placed it to his ear, as he was walking towards the street. I could not hear him say anything. However, very quickly, he put the phone down in front of him hitting the key and turning it off.

Looking back at the man from immigration, he wanted me to pay attention to a particular Immigration Form he was pointing at. At the same time, he was holding a ballpoint pen, for me to sign my name with. He was in a hurry! Taking the pen, I pointed and said "Right there?" He said "Yes" and held the paper down with both hands as I signed my name.

While signing my name, inside my upper peripheral vision, "My Hun" was leaning forward intently. Obviously, she wanted the situation to end quickly.

The man from immigration, also obviously wanted the situation to end quickly. Because he immediately took out the second form for me to sign and started flipping through pages, explaining as he went.

Again, watching him flip through pages, I eased my head to the right and focused on the beautiful woman that had been sitting with "Osama bin Laden," consciously putting as much about her to memory as I was capable of. The whole time consciously captivated by her beauty and realizing that her dress looked like something from Hawaii, many colors, with blue being the predominate color.

Looking back, I wonder if they were planning a trip or returning from one! Was it Hawaii?

At the same time, I was still aware that the bodyguard was standing just a few feet behind me with a death stare on the back of my head. He was obviously ready, to physically confront me violently, at a moment's notice.

Again, moving my head to the left and straining to look over my shoulder and out through the corner of my eye, I watched as "Osama bin Laden", walked between the cars at the edge of the street. And while looking north dialed his cell phone, then put it to his ear and said a very short comment while looking up the street in a northerly direction! He then hits the button to hang up the phone, turned, and began walking back towards the bodyguard.

While focusing my peripheral vision on the two women that had been sitting with "Osama bin Laden," I continued to listen to the man from immigration and noticed that my wife kept her eyes focused on the papers he was pointing at.

Being as wired to the situation as I was, I knew immediately when the bodyguard turned and walked with "Osama bin Laden" toward the street. Turning my head to the far left, while looking out of the corner of my left eye, I watched as both "Osama bin Laden" and his bodyguard walked in between two cars out into the middle of the street, both turned and looked to the north, with very concerned looks on their faces.

Again, just as before, "Osama bin Laden" dialed a number on his cell phone, put it up to his ear, and waited for only a few seconds. That made what he said a very short comment!

I can only imagine what he was saying! I know what I would have said if I was him, which is "I have been recognized; come get me now!"

Then he pushed the cell phone off button. This time he turned and looked at the bodyguard. The bodyguard's lips moved, then "Osama bin Laden's" lips moved saying something back. It was not in English, as I could not read their lips! The whole time, the man from immigration was talking.

It was obvious that he, "Osama bin Laden", wanted the person or person's on the other end of the phone, to get there to him immediately! It was urgent, and he was urgently concerned that they were taking so long to respond!

I knew in my heart of hearts, that the traffic was very bad in the direction he was looking and that whoever it was, that he was calling, most likely was on the other side of the river. Which meant that they would have to cross the bridge in bumper to bumper, slow moving traffic? I felt confidently safe, for the moment!

As times goes by in this eyewitness account, you will come to realize, that my actions paid off. And the entire Philippines were saved from a devastating internal war. As the entire Philippine Military had to go through a loyalty test after my SALUTE REPORT and a coup was put down! Even the political murders have almost completely stopped!

They both then turned and started walking back toward us. Once they were half way back, "Osama bin Laden" stops, while the bodyguard continued to walk until he was again standing directly behind me.

At this moment, I notice to the right of my peripheral vision that the women with "Osama bin Laden" still at the table had their eyes fixed directly on his face. He was out of my view, but I sensed that he was communicating with them.

At that moment, the man from immigration handed me the ballpoint pen again and held the paper with two hands pressing it on the table while I signed my name and signature.

"My Hun" pulled out of her pocket book, the remainder of the 37,000 pesos and counted it out to him while he focused his attention on the money in her hands.

The whole universe seemed to quiet down and focus in on the "money", she was counting as if with tunnel vision.

The hair on the back of my neck told me that "Osama bin Laden" was also counting the money!

Next, once the man from immigration had the money in his hands and the forms stuffed back into his briefcase. We all three stood up. I reached to shake his hand, smiling and saying thank you. "My Hun" did the same, as we pushed our chairs up toward the table and stepped away. The whole time we did this, I was conscious of the bodyguard acting man and his death stair on the back and now the side of my head.

As we turned, and without pausing, I focused my vision directly at "Osama bin Laden", then as I continued to turn my head. I looked at all four of them, using my entire visual spectrum as I turned my head and my body to the right, while taking a step in the direction of the street.

As we walked to the street, the man from immigration was walking on the left side of me. This was a perfect opportunity, so while looking at him; I turned my head, looking over my left shoulder to ensure that my wife was following.

Furthermore, to see if "Osama bin Laden" and his bodyguard was not following also!

I was correct, they were not following. In fact, they were standing together watching the 3 of us walk away. The two women were still seated at the table, they had never moved!

When we had gotten to the sidewalk and turned left, walking back towards the office of immigration, I stopped us as we were completely out of view from the coffee shop and "Osama bin Laden". I wanted to question the man from immigration about the situation that had just taken place.

In my mind, he obviously knew what had happened behind my back, and he too was acting like he did not know what was going on!

He stopped and looked at me as I said; "Did you see how those people were acting?" He looked at me and said; "Who, the Islamic people? Don't worry about them!"

I then said, sternly; "Are you kidding me? Did you see the way they were acting? They wanted to kill me!" He said "No, don't worry about them!" I then said, again sternly; "You need to pay attention to what's going on around you, you're going to get me killed!"

Slightly, he turned his head to the left and directed his gaze across the street as if not wanting to look me in the eye while saying; "No, don't worry about them!"

I was disgusted, and I turned to walk across the street while stating; in a Derogative manner, "Yeah, right!"

As my wife and I got into the car, I cranked the engine, and then turned on the air conditioner. Then I turned to My Hun and said; "Did you see how those people were acting?

Did you see the look on that man's face as he stood behind me?" She stated; "Yes, I saw it. I saw all of it!"

I said to her, "Do you have our camera? She said "I have a camera on my cell phone."

I said, "Get it ready. I want their picture, get it ready now!"

As I watched her push buttons on the phone I asked "Is it ready? Are you ready? We have to hurry; I want you to take their picture as we drive by!"

She then said "Yes, I'm ready!"

As I put the car into first gear and pulled out into the street. Slowly, I drove by, while saying get ready, get ready. My Hun says, "I'm ready I'm ready!"

As I drove by very slowly, I stated; "Take the picture," at the exact moment, we were in between two cars, and I could see the bodyguard still standing where he had been standing alone when we left. He was now staring straight at us! There just behind him up close to the table was "Osama bin Laden." His head turning away toward the two women, who were still sitting at the table!

I continued to say, "Take the picture take it now! Take a picture; take it now!" My wife says; "I got it!" Then she said again, "I got it!"

As we drove on, I turned and looked out the back of the window and watched as the bodyguard continued to watch as we were driving away.

Did he get our car tag number?

As we drove further I needed to make a left turn, and I asked twice; "Did you get the picture? Did you get the picture?"

Repeating myself, I said, "Are you sure, are you sure!"
She said "Yes, I got it, here look!" As I turned and looked at the small screen on her cell phone, I could see the bodyguard with his **orange** baseball cap, staring at me with "Osama bin Laden" clearly in the background.

As we drove further, I was tempted to try to go back around and look for a parking spot. So that I could watch and see if they left and maybe get a tag number off of the car.

Just then, I looked down at the gas gauge on our rental car. It showed that we had just barely enough gas to get to a gas station.

I then asked My Hun; "Do you have the number to the US Embassy?" She said "No! I left it at home." I then said, "Call information and get the phone number to the US Embassy." She said, "I don't have a load it's all gone! We used it up sending text messages, to the man from immigration!"

As we drove on, a debate went on in my mind as to whether I should spend the 600 pesos I had in my pocket on gasoline that would get us home or to buy a load to call the US Embassy. While thinking this through, I asked my wife, "What time is it?"

After hitting a few buttons on her cell phone. She looked at the time and said "20 minutes after four." I knew instantly that I was not going to be able to get in contact with anybody at the US Embassy. As they were close to shutting down tight for the day! They did so, at 4:30pm every day of the week. And to top it off, I had firsthand knowledge that 90% of the people working at the US Embassy were not even US citizens and would be of no help at all. Hell, this is the most corrupt country in Asia; they might even be working with "Osama bin Laden" himself!

Just then, we came upon a gas station. I pulled in, stopped and looked at her as I said, "I need to contact an American! I need to let someone know what just happened! Do you know who that was?"

Her eyes got real big, and she looked down as she said, "No!" I stated loudly; "That was "Osama bin Laden!" Did you see how he was acting? He was pacing back and forth afraid that I had recognized him."

She instantly had a blank stare on her face and became totally silent. A moment later she asked "Who is that?"

I explained quickly!

I was now in an impossible situation! Sitting in my car with “My Hun” at a gas station with just enough money to buy gasoline for the four-hour trip back to our home! Where I would need to turn in the rental car! That same day or have to pay the fee for one more day!

I was in a far away land. A place reported on CNN, as the most corrupt country in Asia. The person with me "My Hun" was scared half out of her mind and depending on me to protect her.

I was spiraling through an insane situation!

I needed to make a phone call, but we had no load for the cell phone. Also, I had no phone number to call. It was also closing time for the US Embassy, which was about 2 miles away, through heavy bumper to bumper traffic.

I knew that I would have to get permission from a non-US citizen who might give the information to someone who might be in contact with "Osama bin Laden". My mind started swimming with thoughts that were going in many directions. This was impossible. My instinct told me to turn around and go back and watch them, but my wife's fear also told me, that I should not put her into any unnecessary danger, as she was not prepared.

As uncomfortable as it was I knew that I needed to spend the 600 pesos in my pocket to buy gasoline and head home. Here in the Philippines, gasoline was well over $4.00 a gallon, long before it was in the USA!

After pulling up to the gas pump and spending the 600 pesos. I became concerned that we had been followed as other cars also had pulled up into the gas station and parked without coming all the way up to the gas pumps or getting in line at the gas pumps. They were just sitting there!

Cranking up the car I drove out into the traffic, turned north in the direction we needed to go to get back home. The traffic was heavy, as it was quitting time, and everyone was headed home from work.

The traffic was bumper to bumper in both directions, and I was back to pop-in the clutch, changing between first and second gear every 3 to 4 seconds.

As we inched forward, the hair on the back of my neck was standing up! I sensed that somehow, somewhere back behind me was someone that was following me. Call this paranoia if you wish, I feel the situation justifies my concern!

With the traffic inching forward I became more and more anxious. I needed to be moving! I also needed to be in touch with an American, today!

Suddenly, there was a break in the traffic going in the opposite direction back in the direction I had just come from.

Instantly, I knew I was going to turn and get into that break in the traffic.

I did just that!

Very quickly after I had changed over into the opposite lane by doing an emergency U-turn! The traffic again was grid locked, as I consciously watched the cars that were in the opposite lane now, to see if anyone was staring in my direction. I did not see anyone!

After what seemed like an eternity, the traffic seemed to break loose, and we inched forward moving a little faster! Then, I could shift into third gear and move along steadily.

After a few miles, we drove past the US Embassy, and sure enough it was locked down tight.

We continued on, as I remarked to my wife "I want to find the United States Veterans Administration Medical Clinic," that is supposed to be on the same street as the US Embassy. I ask her to "Help me look for it!"

The traffic was heavy, but steadily moving, and we strained our necks to look at every building for the name "United States Veterans Administration."

I was thinking to myself: Surely, I must have gone past where we would turn left at an intersection and find it.

Turning around we drove back the way we had just come down, looking out the right window of the car over my wife's shoulder and my wife's side of the car. We continue to look for a sign that said, "United States Veterans Administration."

Moving on, we did not find it, and once we came back to the location of the "United States Embassy," I was becoming very frustrated. I was bound and determined that I was going to find an American.

Turning around, I came back onto the highway and started heading south away from the US Embassy in the direction I remembered reading as to where the "United States Veterans Administration Medical Clinic" was located.

Seeming like forever, we drove further and further and further, then suddenly there was a little gray building in comparison to the larger buildings of the area. It had a small sign that said "United States Veterans Administration Medical Clinic."

I pulled up into the parking lot and stopped. My wife and I got out and walked to the door. It was locked! I knocked on the door, nothing happened. I knocked on the door again. Nothing happened! I walked over to the gate and looked inside; there were people sitting at tables looking at me like, there was something wrong with me. They were Filipinos!

At that moment, I heard a buzzer. My wife turned the door handle of the door I'd been knocking on, and it opened. We both walked inside.

I was very frustrated about the situation and said to the security guard, "I have been looking for this building all afternoon." He starts laughing and says its 5:30pm. I then said, "Well, at least I found it now!" He continued to laugh and says "Its 5:30pm. The clinic closed at 4:30pm. You're an hour late! What time was your appointment?"

I did not feel good about the way he acted so, non-professional. A security guard should always be serious!

Too many security guards here in the Philippines, are only janitors with a uniform, given a badge and a gun.

I told him that I did not have an appointment, but that I needed to speak to an American. So I asked; "Is there an

American inside the building?" He answered, "No! Everybody has gone home!"

The place was locked down and would not be open again until 7:30am of the next working day. As he continued to laugh, I knew that I was not going to discuss anything with him further. I turned, walked away and got back into the car, "My Hun" followed!

Once we had gotten back into the car, I just sat there! I was totally exhausted and could not believe the problems I was having. I.e. I had no money in my pocket. I had already taken out my debit card maximum amount allowed for the day, to pay the remainder of my visa. I had to get the rental car back home or have to pay for another day. My gas was tight, and I might not make it back to our apartment. And if this happened, I had no money for a load for the cell phone to contact anyone to come get us.

Totally exhausted, I was going to have to make the four-hour trip through heavy traffic! This was quitting time traffic, and if I remember correctly it was a Thursday afternoon.

Well, it was done, this was it, and we had to go home. I backed the car out of the parking space, drove to the nearest intersection and got into the traffic.

After we had spent an hour and a half creeping through the city of Manila to get back out toward the 4 Lane Hwy, and headed north!

I turned to my wife as I said, "You saw how those people were acting, right?"

She said, "Yes, I saw it!"

We again repeated the same conversation as before!

I said, "You know who that was, right!"

She became immediately silent!

I said, "That was "Osama bin Laden!

You know who "Osama bin Laden" is?"

She looked at me out of the corner of her eye and said "No!"

I said "Osama bin Laden, the world's most wanted terrorists. You remember 911 and the jets flying into the World Trade Center!"

She said "Yes!"

I said "That was him, "Osama bin Laden" the head man of the entire terrorist group, and he was right there at the next table to us! Everyone's looking for him with a beard and turban. Putting out pictures of him wearing the traditional Islamic clothing! This man had a short haircut; almost a military high and tight haircut. He had his face completely shaved! His skin looked like he had been to a dermatologist and was getting body scrubs. Even his eyebrows looked like they were trim back, and he has obviously been living very good.

"My Hun" then turned to me and looked at me with a blank stare on her face. She was very quiet. I then said, "I've got to contact President Bush. I need to speak to an American. I need to call somebody and let them know that "Osama bin Laden" is here in the Philippines."

"My Hun" looked at me with tears in her eyes very seriously and said in broken English, "I divorce you!"

I replied with a loudly surprised "WHAT?"

The rest of the trip home, we did not speak. There was nothing to say! We were both tired; it had been an exhausting day. My Honey was obviously frightened to death!

Once we got home, and inside the house. My wife was messing with the phone and taking it apart. I said, "Let me see the pictures. I want to get them developed!"

She continued to take the phone apart and said "It's not working!"

I did not believe her and said "No, just leave it alone. I need those pictures!"

She continued to pull it apart taking off the battery then pulling out the system card, as I questioned, "What are you doing?"

She stayed quiet as she continued to pull the system card out. Look at it and then put it back into place. I continued to pressure her to let me see the pictures, and continued to state that I need to have the pictures developed; she stayed quiet!

Once she had the cell phone back together. I said "Let me see the pictures!"

She turned toward me and said "See, it's not working!"

I looked at the screen, it was blank!

Caution, engulfed me! I was at a crossroad! I wanted to explode and scream at her, but I held my tongue. The rest of the night was tense, as we remained quiet and watch the evening, late news on cable TV, which is the morning news in the USA, FOX and CNN!

Over the next few days, I was easily frustrated, wanting to pull my hair out as I tried to think of a way to get control of the situation. I had to work covertly inside my own home! As you might guess, that is, how I have written this book!

We had never been able to afford a land line phone for the house. And my Internet service provider had turned off my Internet connection because of nonpayment! Each time I went to an Internet café, it was standing room only because of the children who were out of school for their Christmas vacation and spending their money at the Internet café, playing online games like "World of Warcraft".

From time to time, I would question "My Hun" about the photographs, and she continually held her ground saying that the cell phone was not working. She could not get it to work. She did not know why! When I searched for it, I could not find it! Surely, she had it hidden somewhere inside our apartment!

As the following days went by, we needed to go for Christmas shopping, but had very little money to spend. Since we have had to pay such a large amount this month at immigration for my and Bubby's visa! Along with renting the car, an extra day, and the price of gasoline, being well over four dollars a gallon. Buying "My Little Angels" milk and diapers, and given "Bubby" his weekly allowance. All of this compounded the stress and strains of not knowing what was going on with my "Social Security Disability Compensation."

We slowly went through one of the poorest Christmas seasons of my entire life. This also took its toll on Bubby! Because he had grown up as an only child, always having an abundance of Christmas presents.

This time, it was one small inexpensive gift!

January brought another payment from the Veterans Administration Disability Compensation. And, after being so cash starved at Christmas, we went out and spent money on Christmas presents that we could not afford during Dec 2007 and had another Christmas. This time, "Bubby" did a little better!

Then one night while we were walking to the night market "Much like a Flea market, in the USA!" over at Subic Bay Freeport. A man dressed in solid black battle dress uniform with no headgear. Walked directly up to me in the crowd. As he passes me, he looks me in the eye and says,

"Follow, you die!"

Concerned, I thought for an instant, "Did I hear him correctly?" Turning my head to the right, looking over my shoulder at the crowd of people walking along with us! The entire street was solid with people walking towards the night market. The man in black instantly disappeared in the crowd of people walking along with us. It was at night with no street lights on this bridge. Hundreds of people walking shoulder to shoulder, he just disappeared among them.

Continuing on, walking with the crowd and being taller than the people around me. I could see two men in black berets. They had, sub machine guns, and were holding them with both hands and <u>with their finger on the trigger</u>. In addition, wearing solid black battle dress uniforms!

As they both looked at me, their faces angry, I felt as if they recognized me. It was as if they were with the man in black, who had just passed me saying;

"Follow, you die!"

Very concerned, moving forward with the crowd, I attempted to act like they were of no concern to me. As we passed them, we turned to the right walking on toward the night market.

Just as the study manuals from Lion Investigation Academy had taught me… I made a firm conscious effort to put this to memory. Somehow I would find a way to report it to an American in authority! Knowing I would have to do it without my wife's knowledge, as she would start shaking from fear!

As you might assume, being from the Philippines, she is very small for a grown woman, and is easily intimidated by people much larger than she is!

Chapter 13

SALUTE REPORT

For the rest of January 2008, I waited patiently. Each time I found myself alone, I went straight to my computer and began working on a folder filled with notes. Because, as the days went by, more events happened that I would soon be reporting about.

As soon as I received my "Veterans Administration Disability Compensation" check deposited directly into my bank account, I would catch up the payment for my Internet access. Find out why my disability compensation from the "Social Security Administration" had not been turned back on and raise hell about it!

At the same time, I would start reporting about my encounter with "Osama bin Laden" and the days to follow.

What you will soon read, is the exact e-mail correspondence's concerning just that!

So let's move on and get closer to it!

"My Little Angel" and her mother had moved in with us temporarily, right after the New Year. It was a hard financial time for all of us, so I had invited "My Little Angel's" mother to come stay with us and to bring "My Little Angel." That way I could more easily take care of her, by providing for her needs without travel expenses.

One day after being cooped up in the apartment for several days. I suggested that we take "My Little Angel," over to the city park. To introduce her to the many playground activities such as swings and slides!

She had just had her second birthday the month before, and her only experience at a playground was inside McDonald's. The city park playground, had swings and slides that were very large, actually too big for her. She spent her time running back and forth between them.

After an hour and a half, My Hun and her sister wanted to take "My Little Angel," further into the city and look around. I did not have the same aspirations and suggested that they go along without me, so that I could walk back to the apartment.

Walking out to the sidewalk along side of the street, we said our goodbyes, and I waved as they were walking away.

I then turned to walk back towards the apartment, and as I came to the first side street. I noticed, up ahead of me, just around the curve, some Filipino men in camouflage military uniforms all standing around a truck.

Immediately, one of the men with them noticed me walking in their direction and turned sideways, while saying something to them. I could easily read his lips and hand and arm movements. He said, "Okay, there he is, let's go, get in the truck; let's go!

Immediately, the men standing around the truck all looked in my direction and began to climb into the back of the truck. They then drove away.

Again, I made a conscious effort to memorize the situation and immediately upon returning home, went to my computer and to my notes. Taking special notice as to what had just taken place.

Furthermore, on several different occasions a man and one-time a woman, and then another time another man came by my apartment over a short two weeks or so period. With a clipboard, pen and paper, each one would ask how many people live here, who they are, and who if anyone was making any money or had a job. Again, I put this in my notes on my computer. For I would be reporting about it as soon as I could get my Internet access turned back on!

Further, one night while My Hun and I, were walking to the movie theater! As we entered Subic Bay, My Hun was stopped and carded along with many other people. I found this to be very unusual as it had never happened before.

I was totally suspicious of this, and you will soon find that my suspicions were 100% accurate!

Also, what turned out to be one of my neighbors, a crippled gray-haired man, would sit and watch our apartment all day long, for many weeks! I never questioned him, but felt as though he was reporting, my activity, to someone!

Now that my Internet accessibility was turned back on, and they had come by to fix the antenna, I wanted some answers about my "Social Security Disability Compensation," and an American in authority to know that I had walked up on "Osama bin Laden," in the Philippines.

Read on, as you see copies of my emails and the replies to my messages:

Also, from time to time as you go forward in this eyewitness account, you may feel as though something is missing from the order of email copies. It is because of emails being removed from my AOL account without my permission and then a complete system shutdown of my computer.

It was down for many days while I tried, at home myself, to get it back up and running. I was not successful and had to take it to a computer repair shop. There it was necessary for a complete new motherboard and hard drive to be installed.

They wanted to format my hard drive, but I refused. Knowing that somehow someway, I would recover the notes and copied emails on it! Which I have been able to do while writing this book, with limited success!

As you might imagine, it was a stressful time with my 100% service connected PTSD escalating higher and higher. I was fully prepared to RAT on anybody and everybody about anything!

Below, is a copy of the very first email, I sent to President Bush concerning my encounter with "Osama bin Laden!" Again, my email address and name are altered to protect my identity. As you can see, I started the email with a copy of the email, I had sent to "U.S. Senator Mel Martinez," and then ended it with my "SALUTE REPORT" concerning my encounter with "Osama bin Laden!" I did this in hopes of the highest impact possible!

Furthermore, with my imagination full of CIA, "Men in Black" knocking on my door. While carrying in their hands a long list of questions; concerning my encounter with "Osama bin Laden." (Surely they would have questions!) Even to the extent of the sky's filling with members of the 82nd Airborne,

United States Army! Along with "Magnificent American Bold Eagles," with their wings spread wide, screeching a warrior's cry.

<u>You now have an appropriate perspective</u> and can read the first email, I sent to President Bush concerning my encounter with "Osama bin Laden".

Ps, The day I sent it, I only had a few minutes, as "My Hun" had walked to the store down the street. So I must work fast and send it now, right now!

I had not been able to locate the immigration form copies, so that I could add in the date. However, I knew it was the first part of Dec 2007 and figured once they sent an investigator to question me, I could pull the forms out that day; which would allow me to update the date to Dec 11th? The investigator never came, and no one has ever asked! Mark it down, so that you have it while you do your investigation of the events to follow, even unto this very day!

Now, take notice of the colors in the SALUTE REPORT and remember the colors of the prisoners in Guantanamo Bay Cuba!

From: [redacted]
To: comments@whitehouse.gov
Sent: 2/4/2008 6:32:44 P.M. Taipei Standard Time
Subj: Purple Heart, Permanently Unemployable!

Dear Kind and Understanding President Bush, I need your help! Please see copy of email below....

U.S. Senator Mel Martinez, Sir! Can you get up off your desk chair, walk 10 steps, and ask two questions? Walk back

[redacted]

[redacted]

During the phone call: I said who I was, and she said that she was waiting for my medical files from the " [redacted] VA Out-Patient Clinic", in [redacted] Fla; I said "Oh No, that will take 6 months!" She then said "yeah and laughed out loud." I did not think it was funny, at all...

I then told her that I had been feeling sick and wanted to know if I could go to a Doctor and use my Medicare Health Insurance (Part A & B) Effective date 4-1-94. She explained that if I was in the United States of America, I could, but because I was outside of the United States, I could not use it.

This let me know, that she knows, that a "Social Security Administrative Mistake," was made on my "Disability Compensation." But she as the "Claims Examiner", was not bound to turn back on my "Disability Compensation," for another 6 months [When she had received the records from Ft Myers Fla. "VA Outpatient Clinic".]

Sir, over the last decade, each time I received a form from the SSA, I filled it out and in doing so, informed the SSA, that the VA had all my medical records. Knowing this, I can honestly say, "They have not been doing their job correctly and should be fired, today!"

And as for Social Security "Claims Examiner", [redacted]," at 02-301 2000 ext: 2459 Manila Philippines. If she does not have enough sense to get up off her desk chair, walk the 10 steps to the VA (Door) office and ask them if their records show me as "permanently unemployable" sense 1991. Then she needs to be fired also.

Further, sense the people in your Orlando office cannot send a simple email to me at "[redacted]", like, I have asked them to do, each time I have sent them a Privacy Act Form and update letter's. And if they cannot speed up the "Claims Examiner", [redacted]," ! Then there wasting tax payer money and need to find another job..... If not, they're going to earn you a bad name in our country......

Copy repaired for Resident Bush, CNN and Anderson Copper 360, respectfully.......

[redacted]

Purple Heart [redacted] reporting.

Is the Philippines government helping Osama Ben Laden?

Situation: "Osama Ben Laden Family Members" PHILIPPINES.......

Size: 2 men 2 women. Same family nose, eyes and cheek bones as Osama Ben Laden.

Activity:

While at the immigration office, I walked over to a cafe. Straight out the door of the immigration office, across the street and around the door of a bank. To outside tables with umbrella tops.

As I walk up, 2 men and 2 women with olive skin look straight into my eyes with concern. As I walked to a different table and to the chair, it had water in the seat. Noticing that they were still looking at me, I moved to a chair with my back to them.

I looked to the right and could tell the woman closest to me was watching me out of the corner of her eye as she watched the men with her that were now standing behind me. As I continued turning my head to the right, the younger of the 2 men had a death stare on the back of my head. (It made no sense.)

(The man from immigration came and sat down at the table with me and started showing me papers and getting me to sign them.)

All the time the 2 olive skin men were standing behind me with very concerned looks on their faces. The older of the two, kept pacing around (He could not stand still.) As he dialed his cell phone, he walked out of my hearing rage. Said a few words on the phone, then walked back to the younger man with orange hat. 4 times he walked away and made a short call. The 5th time he and the younger man walked out into the street and looked north as the older man made a short cell phone call. He had a very very concerned look on his face. They then walked back to the women, the whole time unhappy... The body guard acting younger man continued to stand behind me and stare at the back of my head. The older man would never make eye contact with me.... He looked like Osama Ben Laden with no hair.

[Acting very concerned that I was there. Man that looked like Osama Ben Laden without beard (Totally clean shave.) Could not stand still after seeing me. Paced around with cell phone, dialing and making very short comments. Repeats same activity 4 times. Even walks out into street looking north as body guard man in orange hat followed closely on 5th time. And Repeats dealing cell phone and making short comments on phone.]

Man in orange hat stared at the back of my head and into my eyes very anger. Even watched me when I left, and until I was out of site.

Acting as lookout or body guard. Very anger as he looks at the back of my head, into my eyes as I turn to look at him, and as I walk away. Continues to watch me as I turned corner 200 feet away. Even as I entered car and drove away.....

[The 2 women of same age as older man (mid to late 40's) but very very well kept, even beautiful, olive skin, long hair, well dressed but to many colors for Philippines's area.

They would have fit in more, in Hawaii.... Sat keeping me in the corner of their eyes and the older man that looked like Osama Ben Laden in their main view.]

Location: 100 meters in front of immigration office Manila Philippines-Next to immigration office (Small outside cafe), Manila Philippines. Opox - 2 miles from US Embassy.

Uniform: 1 man looks like Osama Ben Laden. Over 6 foot (6'3" to 6'5" tall) 195 lbs, long nose, No beard, olive skin, very short hair cut. (Totally clean shave. Even eye brows look cut.). Regular US street clothes (Very New). Shoes: penny lo-fur type. Solid light orange colored shirt, button up front, collar.
1 man about 6'6" 30 to 35 years of age 225 lbs, (bright orange base ball cap). Jeans, Solid blue T-Shirt. Boots black.
2 women, olive skin, long nose, same as pictures of Osama Ben Laden family. Beautiful long, thin, loose fit, dresses, many colors, but blue stands out the most.... or as darkest color....

Time: 2:30pm, 5 Dec 2007

Equipment: 1 cell phone.

SALUTE REPORT

Situation: Security taken to a higher state!

[I was walking from the park back to my apartment....] As I walk through the city of Olongapo, a truck load of military men in uniform sees me and turn, load their truck and drive to next street... I read the lips of the man in charge as he said, "There he is, ok let's go." "Let's go, get in."

Further, (never before) as my wife and I walk in the front gate to Subic Bay (To see a movie) they check all Ids. When they see mine, they look down as if they do not want me to see their eyes, then let me go through. And as my wife shows her ID, they want her to fill out a form, then when they say to me go ahead, and I say, " I am with my wife, they point to her, saying, you do not need to do that and then let us though.....

There's something wrong with this............!!!!!!!!!

Size: Battalion size

Activity: Stopping everyone and checking ID's

Location: Subic Bay Free Port (SBMA) and inside city of Olongapo, many military lookouts active....

Uniform: Common street clothes, Subic Security uniforms and Military uniforms...

Time: Ongoing, Jan 2007

Equipment: Common to all Security and Military.

I had sent the same email to Vice President Cheney. Although I do not have the copy of the automatic reply from comments@whitehouse.gov, I do have the copy of the automatic reply from vice_president@whitehouse.gov, see copy next:

Date: 02/04/2008 06:36:42 PM Taipei Standard Time
From:vice_president@whitehouse.gov
To: [redacted]
Subj: **(No Subject)**

Thank you for e-mailing Vice President Cheney. Your comments, suggestions and concerns are important to him.

Unfortunately, because of the large volume of e-mail received, the Vice President cannot personally respond to each message. However, members of the Vice President's staff consider and report citizen ideas and concerns.

Please visit the White House web site for the most up-to-date information on Presidential initiatives, current events, and topics of interest to you.

Thank you again for taking the time to write.

I had sent the same message though a form at usa.gov.com, but with only the copy of the email about Senator Mel Martinez. And received this reply....

-----Original Message-----
From: USA.gov1@mail.fedinfo.gov
To: [redacted]
Sent: Tue, 5 Feb 2008 3:00 am
Subject: RE: Social Security and Medicare [T200802040020] [FGE8000] [FGE999]

Thank you for contacting USA.gov.

We have received your message and understand that you are having issues with your Social Security International Office located in the Philippines (Manila).

If you have been unable to resolve an issue with a federal government agency, you can contact your Congressional representative's constituent services office in your district via telephone or mail.

Often, the constituent services division of a Congressional office is able to intercede on their constituents' behalf. A Senator or Representative may be able to assist you by making an inquiry with the agency about your case or asking the agency to review or expedite a decision. Complete details can be found on your Congressman's website, usually in the "Services" section. To locate the website for your Senators or Representatives and a list of guidelines and procedures, please visit the links below.

To access a Senator's website, please visit http // www senate gov/
To locate a Representative's website, please visit http // www house gov/

When filing a complaint with the Social Security Administration (SSA), it is very important that you direct your complaint through the appropriate channels. First, you should file your complaint in writing with the SSA office where your complaint originated.

If the problem cannot be resolved through the originating office, you may file your complaint, in writing, through the appropriate regional SSA office. For a directory of regional offices, please visit the following site: http // www ssa gov / regions /

You may also contact SSA International Operations by phone: http // www ssa gov / foreign / phones html

We hope you find this information helpful.

Regards, USA.gov Citizen Response Team

We cannot accept replies at this e-mail address. If you have further questions, please send us a message through our web form at: http // USA gov / questions

You can also call our National Contact Center at 1-800-FED-INFO (1-800-333-4636) from the USA and Canada or 1-202-208-1576 from elsewhere. We are open Monday through Friday from 8 a.m. to 8 p.m., ET.
Sender : [redacted]
Tracking Number : T200802040020Z656993
Pool : USA.gov - S.S./Medicare
Sent to : USA.gov@mail.fedinfo.gov

Date : 2/4/2008 5:26 AM
Zip Code: [redacted]

U.S. Senator Mel Martinez, Sir ! Can you get up off your desk chair, walk 10 steps, ask two questions? Walk back to your desk chair, set down, open a file, stamp it completed. Move it into the out box of your desk?

How hard can it be?

Well, Social Security "Claims Examiner", [redacted]," at 02-301-2000 ext: 2459 Manila Philippines, can't even do something as simple as that, and should be fired for in competence.....

I have sent her everything (Forms) that I have sent to your Orlando Office.

To: Office of U.S. Senator Mel Martinez
315 East Robinson Street
Landmark Center One, Suite 475
Orlando, Florida 32801

And still she cannot walk the 10 steps from her desk to the front desk of the VA in the same building as her office, and ask 2 questions!

1.)Is veteran [redacted] in the VA system as 100% service connected disability rating? Yes!
2.)Is same veteran rated as permanently unemployable? Yes!

On the morning of Jan 25, 2008; I called the Social Security Administration Office in Manila Philippines at 02-301-2000 extension-2621 and talked to Social Security "Claims Examiner", [redacted]," at 02-301-2000 ext: 2459. To ask for an update to the "Social Security Administrative Mistake," that was made on my "Disability Compensation."
[It's very simple:
They sent me a form to fill out.
I was not at the address they sent the form to.
I did not receive the form and therefore could not fill it out and return it to them!]
During the phone call: I said who I was and she said that she was waiting for my medical files from the "[redacted] VA Out-Patient Clinic", in [redacted] Fla. I said "Oh No that will take 6 months!" She then said "yeah and laughed out loud." I did not think it was funny, at all...
I then told her that I had been feeling sick and wanted to know if I could go to a Doctor and use my Medicare Health Insurance (Part A & B) Effective date 4-1-94. She explained that if I was in the United States of America, I could, but because I was outside of the United States, I could not use it. This let me know, that she knows, that a "Social Security Administrative Mistake," was made on my "Disability Compensation." But she as the "Claims Examiner", was not bound to turn back on my "Disability Compensation," for another 6 months [When she had received the records from [redacted] Fla. "VA Outpatient Clinic".]
Sir, over the last decade, each time I received a form from the SSA, I filled it out and in doing so, informed the SSA, that the VA had all my medical records.
Knowing this, I can honestly say, "They have not been doing there job correctly and should be fired, today!"

And as for Social Security "Claims Examiner", [redacted]," at 02-301-2000 ext: 2459 Manila Philippines. If she does not have enough sense to get up off her desk chair, walk the 10 steps to the VA (Door) office and ask them if there records show me as "permanently unemployable" sense 1991. Then she needs to be fired also.

Further, sense the people in your Orlando office can not send a simple email to me at "[redacted]", like I have asked them to do, each time I have sent them a Privacy Act Form and update letter's. And if they cannot speed up the "Claims Examiner", Patricio Everly," ! Then there wasting tax payer money and need to find another job..... If not, they're going to earn you a bad name in our country......

Copy repaired for President Bush, CNN and Anderson Copper 360, respectfully.......

[redacted]

Date: 02/04/2008 06:26:53 PM Taipei Standard Time
From: USA.gov1@mail.fedinfo.gov
To: [redacted]
Subj: **Auto-Acknowledgement from USA.gov [T200802040020]**

Dear USA.gov Visitor,

Thank you for contacting USA.gov. Our goal is to make it easy for you to communicate with the government.

A member of our Citizen Response Team will respond to you within two business days. We appreciate your patience.

For future reference, your message ID number is T200802040020.

We regret that we are unable to accept replies sent to our mailbox. If you would prefer to contact us by telephone, we encourage you to call our National Contact Center at 1-800-FEDINFO (1-800-333-4636), Monday through Friday from 8 a.m. to 8 p.m. Eastern time.

Regards,
USA.gov Citizen Response Team

In a message dated 2/8/2008 4:49:47 A.M. Taipei Standard Time, Lane_Mahood@martinez.senate.gov writes:
Dear Mr. [redacted]:
I wanted to inform you that I am working on your case for Senator Martinez. I will be in touch with you again via e-mail when I receive a response on your case.
Sincerely,
Lane Mahood
****
Lane Mahood
Senior Constituent Services Representative
U.S. Senator Mel Martinez
315 East Robinson Street
Landmark Center One, Suite 475
Orlando, Florida 32801
Phone: 407-254-2573
Fax: 407-423-0941
http // Martinez senate gov /

*****

Oh yes, now I knew someone had received and read my email to President Bush. And I knew the hunt for "Osama bin Laden" was now focused on the Philippines. I had even hoped that my Social Security Disability Compensation would be turned back on soon. This way I could rent a car, and do some looking around myself!

However, it didn't, and I had to sit at home. Doing just that, I witnessed on the local Philippine news channel, the entire Philippines military undergo a complete nationwide loyalty test. A codetta by military leaders be discovered, and put down.

As the weeks went by, I became angry again concerning my Social Security Disability Compensation. So I continued forward with my enquirers!

From: [redacted] [mailto: [redacted]]
Sent: Friday, February 22, 2008 10:16 AM
To: Mahood, Lane (Martinez)
Subject: Re: Inquiry with Senator Martinez

Sir, I need an update!
[redacted]

*****

I received this in reply, see next:

In a message dated 2/23/2008 3:50:36 A.M. Taipei Standard Time, Lane_Mahood@martinez.senate.gov writes:

Dear Mr. [redacted]:
I am awaiting a response from the Social Security Office of International Operations. I will be in touch with you when I have more information.

Lane Mahood
Senior Constituent Services Representative
U.S. Senator Mel Martinez
Phone: 407-254-2573
That message set my PTSD on fire:

Subj: **Re: Inquiry with Senator Martinez**
Date: 03/03/2008 09:52:59 PM Taipei Standard Time
From: [redacted]
To: Lane_Mahood@martinez.senate.gov
CC: comments@whitehouse.gov,
vice_president@whitehouse.gov,
Charlie.Crist@MyFlorida.com

Sir, I need an update. It does not take social security more than 24 hours to respond to the office of a US Senator.

Also, why in the world would you be trying to contact the Social Security Office of International Operations, when I am a Florida resident? I only came here to the Philippines to visit for a few months before I was to return to my home in [redacted] Florida. There (SSA) sending the form for me to fill out to an old out dated address was what caused the whole problem. And I cannot get up enough money to fly my son and I back to Florida until my social security disability compensation is turned back on. I am stuck here because of a Social Security Administration mistake. And, I can't seem to get you to understand English! Maybe I should speak Spanish! What does "permanently unemployable" mean to you?

When I contacted you a few weeks ago for an update, I had just found out that this same thing happened to a Vet in FT Myers Florida and after a TV reporter worked the case for 1 week, the social security office turned back on his disability compensation. Two days later I was talking to a Vet here that is "permanently unemployable" just like me. And he said that the social security administration did the same thing to him. But after he went to their office and caused a big seen, they turned it back on in 3 days. And he said that I should do the same thing. Of which I am getting ready to do.

If this is not resolved this week, ending on 7 March 2008, I fully intend to get in touch with the TV reporter in FT Myers Florida as soon thereafter as possible. And I will send him whatever, he needs from me to get the job done. (Names, Dates, Phone Numbers, and release forms) Which is what I did with your office months ago?

And to top it off, not having the money for a tank of gas or a phone call, so I could follow "Osama Ben Laden" when I saw him. Allowed him to get away; Another Social Security Administration mistake!

Thank you, [redacted]

Copies sent to President Bush at comments@whitehouse.gov , Vice President Cheney at

vice_president@whitehouse.gov and Governor Crist at Charlie.Crist@MyFlorida.com

*****

Now that I had their attention, I sent the following "SALUTE REPORTS" to as many US Gov emails and website contact forms as I could without "My Hun" walking in and discovering what I was doing.

SALUTE REPORT

Purple Heart [redacted] reporting.
Is the Philippines government helping Osama Ben Laden?
Situation: Osama Ben Laden family members, Philippines.
Size: 2 men 2 women. Same family nose, eyes and cheekbones as Osama bin Laden.
Activity: while at the immigration office, I walked over to a cafe. Straight out the door of the immigration office, across the street and around the door of a bank to outside tables with Umbrella tops…
As I walk up, 2 men and 2 women with olive skin look straight into my eyes with concern. As I walked to a different table and to the chair, it had water in the seat. Noticing that they were still looking at me, I'm moved to a chair with my back to them.
I look to the right and could tell the woman closest to me was watching me out of the corner of her eye as she watched the men with her that were now standing behind me. As I continue turning my head to the right, the younger of the two men had a death stare on the back of my head. It made no sense.
The man from immigration came and sat down at the table with me and started showing me papers and getting me to

sign them. All the time the two olive skinned men were standing behind me with very concerned looks on their faces. The older of the two, kept pacing around, he could not stand still. As he dialed his cell phone, he walked out of my hearing range. Said a few words on the phone, and then walked back to the younger man with orange hat. Four times he walked away and made a short call. The 5th time he and the younger man walked out into the street and look north as the older man made a short cell phone call. He had a very very concerned look on his face. They then walked back to the women, the whole time un happy. The bodyguard acting younger man continued to stand behind me and stare at the back of my head. The older man would never make eye contact with me. He looked like Osama Ben Laden with no hair

UPDATE : man that looked like Osama Ben Laden -- looked as if he has been putting something below his eyes to whiten or lighten the skin.

Acting very concerned that I was there. Man that looked like Osama Ben Laden without beard, totally clean shave. Could not stand still after seeing me. Paced around with cell phone, dialing and making very short comments. Repeats same activity 4 times. Even walks out into the street, looking north, as bodyguard man in orange hat followed closely 5th time. And repeats dialing cell phone and making short comments on phone.

Man in orange hat stared at the back of my head and into my eyes very angry. Even watched me when I left, and until I was out of sight.

Acting as lookout or bodyguard. Very angry as he looks at the back of my head, into my eyes as I turn to look at him, and as I walk away. Continues to watch me as I turned the corner 200 feet away. Even as I entered the car and drove away.

The 2 women of same age as older man mid-to-late 40s, but very very well kept, even beautiful, olive skin, long hair, well-dressed but too many colors for the Philippines area. They would have fit in more, in Hawaii. Sat keeping me in the corner of their eyes and the older man that looked like Osama Ben Laden in their main view.

Location: 100m in front of immigration office Manila Philippines -- next to immigration office small outside Café, Manila Philippines. Approximately -- 2 miles from US Embassy.

Uniform: one-man looks like Osama Ben Laden. Over 6 ft 6" tall 195 pounds, long nose, no beard, olive skin, very short haircut. Totally clean shave. Even eyebrows look cut. Regular US street clothes, very new. Shoes: penny loafer type. Solid light orange colored shirt, button up front, collar.
1 man about 6'6", 30 to 35 years of age 225 pounds, bright orange baseball cap. Jeans, solid blue T-shirt. Boots black.
2 women, olive skin, long nose, same as pictures of Osama Ben Laden family. Beautiful long, thin, loose fit, dresses, many colors, but blue stands out the most.... or as darkest color....

Time: 2:30 PM, 5 Dec 2007

Equipment: 1 cell phone.

<u>Special note:</u>

If you have Google Earth set up on your computer!
You can find the location in which I walked up on "Osama bin Laden".
Go to Manila Philippines.

Put your mouse over 14 degrees 35'37.72" N and 120 degrees 58'30.94" E...
Then zoom in, and you are there....

SALUTE REPORT

Situation: while walking into Subic Bay, a man in black BDU, no headgear, saw me, walked toward me. As he passed me, he looked me in the eye and said, "Follow you die!" As I realized what he said, I turned to see where he had gone, or if he was following me. Too many people were right behind me, walking in same direction as myself. I could not get a fix on the man in black.
As I continued to walk toward the entrance to Subic Bay, two men in black BDU, Black Berets, automatic weapons standing next to each other. Watch me approach, look me in the eye very angry.
Size: three men late 20s to early 30s.
Activity: watching me.
Location: Subic Bay Freeport SBMA.
Uniform: Black Beret, solid black BDU -- military ['['uniforms. Same as old-style OD -- 107 but solid black. Black jungle boots.
Time: 10 January 2008.
Equipment: 2 NPVS-5's, banana clips.

SALUTE REPORT

Situation, Security taken to a higher state!
I was walking from the park back to my apartment. As I walk through the city of Olongapo, a truck load of military men in uniform sees me and turn, load their truck and drive to the

next street. I read the lips of the man in charge as he said, "There he is, let's go. Let's go, get in!"
Further, never before as my wife and I walk in the front gate to Subic Bay, to see a movie, they check all ID's. When they see mine, they look down as if they did not want me to see their eyes, then let me go through. And as my wife shows her ID, they want her to fill out a form, then when they say to me. Go ahead and I say, "I am with my wife, they point to her, saying, you do not need to do that." And then let us through. There's something wrong with this....!!
Size: Battalion size.
Activity: Stopping everyone checking IDs.
Location: Subic Bay Freeport SBMA, and inside city of Olongapo, many military lookouts active.
Uniform: Common Street clothes, Subic Bay security uniforms and military uniforms.
Time: Ongoing, January 2008.
Equipment: Common to all security and military.
SALUTE REPORT

Situation: While talking to the VA on the phone, a Philippine woman demands I bring her a copy of my DD 214.
Size: 1
Activity: Demanding I bring her a copy of my DD 214.
Location: VA Outpatient Clinic, Manila Philippines.
Uniform: Unknown.
Time: Ongoing, January -- February 2007.
Equipment: none.

Situation: 2 men 1 woman. One at the time, come by the house on different days.
Size: 3.

Activity: Ask who and how many live here.
Location: 107D Esteban St, New ILaLim, Olongapo city, Zambales Philippines 2200.
Uniform: street clothes.
Time: 10 December 2007 through 20 December 2007.
Equipment: pen and paper.

*****

The CIA contact form, on their website, will not let you copy and paste a message into the contact form. Because ", ' , cause's it to reject the type. I did not want to have to write the same message many times, and I was running out of time, before I would be discovered and needed to hurry. So I sent the message to President Bush and VP Cheney, asking the person who read it to forward it to the CIA. See copy next:

-----Original Message-----
From: [redacted]
To: comments@whitehouse.gov
Cc: vice_president@whitehouse.gov
Sent: Sun, 2 Mar 2008 4:05 pm
Subject: Purple Heart reporting.

Purple Heart [redacted] reporting. Please forward to CIA as I cannot get there website contact form to work...

Situation: Very Poor Security at Subic Bay Philippines for US Navy Ships:
How I know!
Today my son [redacted] and I were looking for a street with no traffic, so that I could give him a few lessons on how to drive a stick shift (5 speed Toyota). We left the house at

[redacted] city, Philippines and drove to Subic Bay. After driving through a few streets full of traffic, I turned toward the Bay, came across a security road block and turned to the left to avoid it. As I kept driving on, we came to a road with a security check. So I turned right to avoid it. Driving on southward I came to a closed gate and a turn to the right. So I turned right and drove west. Which brought us to a stop sign; I stopped and looked right (North) and seen a road block by Subic Security, so I turned left to avoid it.

Bang........... I'm looking straight at an American Navy Aircraft Carrier, American soldiers are in their street close walking away from the ship and Philippine soldiers are dressed in their New BDU with the **orange** part of their camo uniforms showing brightly in the midday sun.

I turn right (west) at the corner and drive along the street not believing how easy it was to accidentally drive through all the security road blocks. And right beside, within a few feet, of the soldiers.

As we drove on westward we came to a security road block again, so I pulled into a parking spot in front of the ship named "Niagara Falls". As I looked at the security guards at the road block, I could tell that one of them had spotted me, but was not sure what to do.

After a few minutes, we drove out the same way we came in, un-stopped or checked. And went back to what we were doing....

Size: 7

Activity: Not providing secure security.

Location: Subic Bay Philippines

Uniform: Subic Bay Security Uniforms

Time: 2:00 pm Philippine time.

Equipment: 2 trucks, 1 gate, 12 orange traffic cones.

What's up with all the rust on that carrier? Don't we have any paint for that ship?
It makes the US Navy look poor....

*****

I had also sent a copy through the website contact form to Senator McCain with a copy of my SALUTE REPORT sometime early in Feb 2008 and then later about the security problem at SUBIC BAY. His office sent this reply, see copy on next page:

Wow! My "Patriotic Heart" jumped with joy! This 'American Hero" Senator John McCain had appointed someone in his office with instructions to have enough respect to send a "Thank you!" reply to me, an Unwanted, Unappreciated Combat Wounded Veteran, Of The 82nd Airborne, United States Army!

Gee, everyone else treated me, as if I was "Unappreciated" and that they wished I would just go away!

-----Original Message-----
From: senator@mccain.senate.gov
To: [redacted]
Sent: Wed, 5 Mar 2008 12:55 am
Subject: Correspondence from Senator McCain

DATE March 4, 2008

Mr. [redacted]
Dear Mr. [redacted]:

I wanted to take a moment to thank you for your recent correspondence. I appreciate your taking the time to assist me in keeping up to date. I found the material you referenced to be very informative.

I have passed along the information you forwarded to my foreign-relations and defense staff. I am confident that it will be useful. In the future, please do not hesitate to contact me on this or any other matter.

Again, thank you for the information. I certainly hope that you will continue to keep in contact on this and other matters.

Sincerely,
John McCain
United States Senator

JM/ab Please Do Not Reply to this Message

When I had seen on CNN and FOX News that Senator McCain was visiting the Whitehouse after being nominated as the GOP candidate for President, and President Bush **dancing** on the front steps as he waited for Senator McCain to arrive…

I believed President Bush was really dancing because of my "SALUTE REPORT" concerning my encounter with "Osama bin Laden!" and now another wonderful event had taken place, and even just as wonderful, as my "SALUTE REPORT" concerning my encounter with "Osama bin Laden". Which was Senator McCain's being nominated as the GOP candidate for President? I even wondered if they had already captured "Osama bin Laden."

It was a "TOP SECERT" time in our history, when the "War on Terror" had taken a turn in our favor!

When I had seen on CNN and FOX News that Senator McCain had landed in Iraq and was in a closed door meeting; I knew it was about my "SALUTE REPORT" concerning my encounter with "Osama bin Laden." Then when I saw him using the word al- Qaida, and then corrected by Senator Lieberman as he said, "You mean Insurgents!" I knew in my heart of hearts that the main topic of the closed door meeting was about my "SALUTE REPORT" concerning my encounter with "Osama bin Laden." Or even that they had already gotten him. Then when the media trashed him for it, I was greatly offended! However, I knew Senator McCain was on a "Top Secret" mission, to pass on the info to our commanders in the field, and it most likely was concerning my SALUTE REPORT.

He would never put himself before his country by explaining why the word al-Qaida was on the tip of his tongue causing him to speak it when he meant to say insurgents.

However, then I thought it strange that Senator Lieberman was so quick to correct him, when it could have gone by unnoticed.

I even wondered if he was the one that leaked the info to Senator Obama. Which is why Senator Obama, went on his sudden vacation to the Caribbean? Where he could easily make international phone calls without the FBI or Homeland Security listening in!

"Senator Obama" was obviously shaken up about "Senators McCain's" sudden trip to Iraq, and wanted to know what was going on. So tell me, who told him about my "SALUTE REPORT" and who do you think he called while in the Caribbean? Was it the King of Saudi Arabia? Was it someone in Kenya? Was it "Osama bin Laden?" Was it "The President of the Philippines?"

Why did he have to go to the Caribbean just to make phone calls? Now that you know the truth as to why Senator McCain went to Iraq, you also know the truth as to why Senator Obama went on a sudden, unplanned vacation to the Caribbean! It was to make overseas phone calls that the Whitehouse Staff and Homeland Security could not overhear, by law! It was all "Cloak and Dagger."

By this time, I was back at it with Lane Mahood, see copy next:

In a message dated 3/8/2008 1:33:35 A.M. Taipei Standard Time, Lane_Mahood@martinez.senate.gov writes:

Mr. [redacted]:
I must be in contact with the Office of International Operations because you are a Social Security recipient who is living outside of the United States. I am still awaiting a

response to my inquiry, but I followed up with them today to let them know the urgency of the situation.

Lane Mahood
Senior Constituent Services Representative
U.S. Senator Mel Martinez
Phone: 407-254-2573

As you might imagine, I had had enough of their foot dragging. I needed results, right here right now. It was getting close to a year after my Social Security Disability Compensation had been turned off, and I needed it back.

Subj: **Re: Inquiry with Senator Martinez**
Date: 03/08/2008 06:44:09 AM Taipei Standard Time
From: [redacted]
To: Lane_Mahood@martinez.senate.gov
CC: comments@whitehouse.gov, vice_president@whitehouse.gov, Charlie.Crist@MyFlorida.com

Dear Lane Mahood,

You mean trapped outside the United States! I would have returned to Florida months ago if they had not turned off my disability compensation. And with the way the dollar is going down down down, supporting my wife, my son and myself, is getting harder and harder each month. I don't want to live in a garbage dump like so many people here have to do. I deserve better! And my dependents do also...

The Philippines Immigration Office made me pay them $35,000 peso's. That's right, "Thirty Five Thousand! And next month I will have to pay that same amount again for my son [redacted]. Just 3 days ago I asked him [[redacted]] to go to the office at the University of Saint Lewis in the Philippines and to the VA office to check on him getting started with college sense we're trapped here. The VA told him that they only pay $400 toward his college each month, if he goes to the University of Saint Lewis in the Philippines. What the HELL, things just keep getting worse and worse.....

And let me tell you something!

You work in the office of a "US Senator", when another government office of the United States Government does not respect you enough to respond to your inquiry within 24 hours! Find your way to their desk, stand in front of it, slam both of your fists "knuckle's first," into their desk top, as you lean toward them, and look them strait in the eye, raising your voice. Demanding, the respect that the "Office of United States Senator Martinez" deserves... If you do not, they will continue to disrespect you, and the "Office of United States Senator Martinez", by not giving timely answers, which allow you to do your job. Which result in the issue dragging on and on, unresolved?

Purple Heart [redacted]
Coples sent to President Bush at comments@whitehouse.gov , Vice President Cheney at vice_president@whitehouse.gov and Governor Crist at Charlie.Crist@MyFlorida.com

*****

I went through the next few weeks wanting to be of help to Senator McCain and our people. So I sent to him a version of what I thought would help him and the American people, see copy on next page:

-----Original Message-----
From: [redacted]
To: senator@mccain.senate.gov
Sent: Mon, 24 Mar 2008 11:32 pm
*Subject: From Purple Heart *****

Dear Purple Heart, Senator John McCain.

Kind and Understanding Sir; I do hope that you will be our next President. So I have taken this idea that has been floating inside my mind for a few years now and put it into an "email" for you to read. If needed, I can send a copy though US Mail.

It is not my intention to tell you what to do, but to pass the ball to you, so that you may be the quarterback, taking the ball and running with it...

It has to do with "Emergency Management Program's" and the possible benefits of uniting all Americans around a "Just Cause" called a "Emergency Mass Migration City."

If you find that this idea of an "Emergency Mass Migration City" could be useful to you and our people! Then please feel free to take this idea as your own, build on it, put it before the people of our great country and move forward with it as you feel might best serve you and our people......

You will find that this plan, as is, is short and not completely finished. It needs a "Think Tank" filled with minds that can conceive the more in-depth possibilities of an "Emergency Mass Migration" [Like what happened during "Hurricane Katrina"] and the benefits of an "Emergency Mass Migration City."

The purpose of sending you this plan is to save lives by helping "Elect Senator John McCain As The Next US President", by uniting the citizens of the United States of America in a "Just Cause" plan of action that will diminish the future effects of another "Katrina Type" catastrophe. Creating economical growth, providing jobs and the increase in the value of the US Dollar!

[Combined with a winning strategy on the "War On Terror", this can work, catapulting "Senator John McCain" into the "White House Oval Office" as "Commander And Chief!" A LION to lead our LIONS…

Further, it is with great pride, as I look forward to that day in the future when I get to cast my vote in your favor.

Sir, I send this to you in happiness and with positive thoughts. The ball is in your court, take charge.

Sincerely,
Purple Heart, [redacted]
[redacted]

United Citizen's Anti Katrina Plan

A "JUST CAUSE"!
United We Stand, For The People, By The People!
Arthur: Purple Heart [redacted]

Sections of This Plan Include:
Terms:
Overviews of "United Citizen's Anti Katrina Plan."
Usefulness Of Plan
General Resources

Terms:
Just Cause: A justifiable cause that unites all citizens.
Anti Katrina Migration or Emergency Mass Migration: A major "US City" size, mass migration, brought on by

immediate peril on a mass level. Such as the result's brought about by the warnings put out as "Hurricane Katrina" approached and then the aftermath.

Even a dirty bomb could cause something like this! An "Emergency Mass Migration"! It has already happened and can happen again in our life time!

Overviews of "United Citizen's Anti Katrina Plan…"

What is the "United Citizen's Anti Katrina Plan"?

A plan to build a "Emergency Mass Migration City."
A Plan that we can come together and agree on…
A plan that we can work together on…
A plan that can save lives…
A plan that can create jobs…
A plan "For the People by The People..."

We the people, "Of Our Great Country," can unite and prepare for an "Emergency Mass Migration"
to a, pre-built "Emergency Mass Migration City."

Example Plans:

Plan A. A "United States Military Base" that is no longer needed as a Military Base and then turned into an "Emergency Mass Migration City." There-by, left empty until an "Anti Katrina Migration" or "Emergency Mass Migration" is implemented by the US President, FEMA and or Homeland Security.

Left empty until needed [only security and maintenance personnel on site 24/7 - 365].

Note: An inter city could migrate within 24 to 48 hours to a "Emergency Migration City." Opening Banks, ATM's, Schools, Shopping Centers, Hospitals, City Government Offices, Emergency Services, Airports, Law Enforcement Offices, Jails/Prisons, Animal Hospitals, with completely furnished housing ready for citizens to move into at a moment's notice… (Not temporary type structures, but permanent structures that will last 100 years or more.

Naturally within a short time, citizens would begin emptying the "Emergency Mass Migration City" after their home city has been deemed fit to live in once again. / Leaving us ready again and again for future mammoth size catastrophes and leaving a place for our "Grand Children" to go in an emergency, long after we are gone. (A "Green City" type place, so that our people can continue their day to day life's much sooner than from a temporary shelter...)

Even Wal-Marts, McDonald's, Corner Store, Mom and Pop, and Burger King type businesses could be up and running over night by a phone call to their suppliers with the address for their next inventory/supply shipment.

Managers could give the location (place of business), to employee's, to report to work at, the next day following the "Emergency Mass Migration". The same could be done for schools "Teachers and Students".. City Gov and so on...]

Plan B. Build a "Emergency Mass Migration City". In an area that does not take up farm land or other natural resources. Like say, a desert! Complete with everything a city has, i.e.. Baseball fields, parks, movie theaters, restaurants and so on, same as in the notes of Plan A, left empty until an "Anti Katrina Migration" or "Emergency Mass Migration" is implemented by the US President, FEMA and or Homeland Security. With, only security and maintenance personnel on site 24/7 - 365 days a year!

Plan C. Build 3 to 5 "Emergency Mass Migration City's", strategically positioned across the continental US. With another in Alaska, left empty until an "Anti Katrina Migration" or "Emergency Mass Migration" is implemented by the US President, FEMA and or Homeland Security.

Plan D: Find 5 very small cities / towns inside the Continental US, like Berlin Ga. And build "Emergency Mass Migration City's" the size of the city named Valdosta Ga. around it. Leaving the "City Of Berlin" in-tacked, in the center, as the "Heart" of the "Emergency Mass Migration City"... This way, the 5 "Emergency Mass Migration City's" inside the "Continental US" could affectionately be called "The City of Heart's" location's 1, 2, 3, 4 and 5. Alternatively, "The City's With Hearts"… Even "Our City's With Hearts"… Alternatively, even, in which state it is put and then named after that state. Example: In Texas, " The Heart of Texas", In Florida, "The Heart Of Florida." and so on...

Setting the example for all countries to follow... i.e. Mexico and all of Central and South America and , EU, Russia and China and so forth.. Creating economical growth worldwide... Note: The UN could sponsor a number of "Emergency Cities" around the 3rd world, like say Africa or India that is not controlled by any government. However, by UN only....

Plus: If Saudi Arabia can build a complete city that is left empty, except for 2 weeks out of each year, then we can build many..

Special Note: Our Homeless Veterans and other needy citizens could be given housing and jobs maintaining the empty "Emergency Mass Migration City's" on a temporary basis. Giving them the chance to hold their head up while they work toward a brighter future... Example: A veteran could be given the job of operating a "Street Sweeper Machine" inside one of the "Emergency Mass Migration City's". Gaining experience and then receiving a VA loan to purchase a "Street Sweeper Machine" and obtaining a contract with a city government to clean their streets late at night.

Funding the "United Citizen's Anti Katrina Plan" Mandatory or Voluntary.

a. Voluntary - All employed citizens/non-citizens can and should have 1 to 5 dollars taken from their salaries by their employer's and sent in along with their Social Security taxes once a month on a voluntary basis. 10 million employed citizens/non-citizens sending in 2 dollars per month… Would be 20 million dollars monthly toward the "Emergency Mass Migration City's"…

b. Mandatory - All employed citizens/non-citizen's will have 1 dollar taking from their salaries by their employer's and sent in along with their Social Security tax's once a week on a mandatory basis with a voluntary 2 - 3 - 4 or 5 dollars added, and tax deductible... (The math looks good on this one also.)

c. Mandatory or Voluntary - Same as above but with a scenario like the "War Bonds of WWII ", where each contribution is repaid with interest like a US Savings Bond. It could be called something like, "The "United Citizen's Anti Katrina Plan" Savings Bond".

Usefulness of Plan

"Insuring all construction material" is made in the "USA" and "all contracts" are with "small or family owned businesses", this would catapult economical growth, providing jobs and the increase in the value of the US Dollar.

Uniting the Citizens of the United States of America in a [Just Cause] plan of action, that will diminish the future effects of another "Katrina Type" catastrophe. By providing "Emergency Mass Migration City's" that will be used on a 2 week, 2 month or even a 2 year basis only.

There are certain skills to have when conducting mammoth size projects, such as building and maintaining "Emergency Mass Migration City's". It is best to have the US President, form a team of planners.

General Resources

"The People of the United States" are the greatest resource the world has ever known. United, they have undertaken "Mammoth Size Projects" in the past and can once again.

We The People, "Can And Will Stand United, In An Anti Katrina Plan".

Further Notes:

The US Military could use the surrounding areas of the "Emergency Mass Migration City's" as live mission training areas. Gaining mission experience in "Homeland Security" and protecting the "Emergency Mass Migration City's" at the same time.....

Wow, bang!

I started getting suspicious emails and sent them to the office of Senator McCain and President Bush, see copy on next page:

Subj: **Fwd: Can I Trust you!.**
Date: 03/26/2008 06:21:03 AM Taipei Standard Time
From: [redacted]
To:comments@whitehouse.gov

Dear Sir,
This looks like something that the FBI should look into.

Purple Heart [redacted]

From: r.johnson@shakara.biz
Sent: 3/26/2008 12:52:53 A.M. Taipei Standard Time
Subj: Can I Trust you!.

Hello,

May i introduce myself as Randy L. Johnson, 34, Army Sergeant 1st Class, an American soldier with Swiss background, serving in the military with the army's 3rd infantry division now a casualty in Iraq.

With a very desperate need for assistance, I have summed up courage to contact you. I found your contact particulars in an address journal. I am seeking your kind assistance to move the sum of (US$50 M) to you pending our arrival to for sharing.

Some money in various currencies was discovered in barrels at a farmhouse near one of Saddam's old palaces in Tikrit-Iraq during a rescue operation, and it was agreed by Staff Sgt. Ryan D. Maseth and I that some part of this money to be shared among both of us before his demise.

See link below for more information about the origin of the money: http://www.voanews.com/burmese/archive/2003-04/a-2003-04-20-1-1.cfm

The above figure was given to me as my share, and to conceal this kind of money I need someone outside Iraq, so with the help of a British contact working here and his office enjoy some immunity, I was able to get the package out to a safe location entirely out of trouble spot. He does not know the real contents of the package.

If you are willing to assist me in this venture you are advised to get back to me with your complete contact information which will be forwarded to the diplomat who will be carrying the said consignment to you for safekeeping pending our arrival as you have fourty percent of the total fund.

Get back to me with your contact information asap

Thank you

Sincerely,

Sgt Randy L. Johnson

E-Mail: r.johnson@rescueteam.com

E-Mail: sgt-johnson@live.com

Next, is my return email correspondence from Sen. McCain about my message concerning the" United Citizen's Anti Katrina Plan." It did my heart good, to receive correspondence from this great man!

-----Original Message-----
From: senator@mccain.senate.gov
To: [redacted]
Sent: Mon, 31 Mar 2008 10:54 pm
Subject: Correspondence from Senator McCain
DATE March 31, 2008

Mr. [redacted]
Dear Mr. [redacted]:

Thank you for taking the time to contact me at my U.S. Senate office.

Because your concerns are campaign-related, your inquiry has been forwarded to the appropriate campaign office.

In the future, should you have any questions or comments concerning issues or legislation before the Senate, please do not hesitate to contact me in my Washington, DC, office.

Sincerely,
John McCain
United States Senator
JM/

Please Do Not Reply to this Message

You can see it was a very heartfelt message that I sent to Sen. McCain, about the "Emergency Mass Migration Cities."

And after receiving the correspondence about it, my heart filled with joy.

I don't know why it was never brought up! However, just like you, I watched as Senator McCain said, "I would rather lose an election, then lose a war!"

Maybe something had gone on behind closed doors!

If this two and two equals four, did Senator McCain agree not to bring up my "SALUTE REPORT" and the reason he went to Iraq? Which would hurt the US military's "War on Terror" and kill Osama's chance to win the Presidency?

When you add two and two on this, do you get the same four as I do?

Further, I still believe in the idea of "Emergency Mass Migration Cities."

With the disasters that happen naturally, every single year. From hurricanes to tornadoes, and floods, from massive melting snow runoff! Alternatively, even the horror of a terrorist dirty bomb... Not to mention the mass deaths that could be brought by a sudden massive solar flare.

It's time we get busy and prepare a place that can handle a mass emergency migration.

It is possible, and if you think about the amount of jobs it would create. It's a very good idea that we should start talking about immediately.

The people of America have a history of uniting around a "Just Cause." The "United Citizen's anti-Katrina Plan" can be that "Just Cause".

What I'm trying to express here on this page is my desire to see the "United Citizen's anti-Katrina Plan" become a reality. Not just a figment of my imagination as to what should be or what could be or what needs to be.

So if you have the ability to organize, along with a desire to be a part of a world changing event. Please start talking about the "United Citizen's anti-Katrina Plan" with your family, your friends, your neighbors, the people you meet on the street. And let's make this happen!

Imagine if you will, the entire city of Charleston South Carolina being given emergency evacuation orders by the government of the United States. With all four lane highways open to "One Way Traffic" headed out of the city, which leads to the nearest emergency mass migration city. Because of an approaching, massive category five hurricanes expected to make landfall at high tide.

The entire population of Charleston South Carolina successfully migrating to a pre-determined emergency mass migration city! With all city government, police departments, fire and emergency departments, schools and hospitals, every single citizen and their families, successfully going about their business and day-to-day lives within 24 to 48 hours after the city of Charleston South Carolina being totally destroyed!

Now think about what happened in New Orleans before, during and after Hurricane Katrina. Do we want that to happen again, ever? I know I don't!

We can all agree that this is a "Just Cause." That needs to be addressed now, not set aside, because we have petty differences and wish to point our fingers at someone else and say that they are to blame for what happened during Hurricane Katrina." That would mean that we all have PTSD, wouldn't it?

If we do not come together and prepare properly, for anything in the future that may cause another Emergency Mass Migration. Such as happened with Hurricane Katrina! Then we have no one to blame but ourselves!

Now think about what would happen if the entire coastline of California was given an emergency evacuation order. Because of an approaching horrific tsunami, capable of destroying everything in its path!

Where would they go? They have to go somewhere!

And what if it were weeks, months or years before their communities were livable again! They would need to move on with their lives, and we would need to help them.

Now think about if we planned ahead and had emergency mass migration cities completed. It could be quite the different scenario.

Even with what's happening in our day-to-day lives, right now! If the entire city we live in, had to be evacuated immediately. No time to prepare! We dare to think or wonder! Where would we go? Where would we take our loved ones?

It was a terrible thing to see what happened during Hurricane Katrina. It's also a terrible thing to think, that we have time to prepare. Time to get ready! But we are redirected away from thinking about this horrible situation, by our day-to-day lives. Allowing it to move to the back of our minds, so that we do not prepare…

Do you believe we can build emergency mass migration cities? And not allow them to be abused, but left empty and ready for an emergency mass migration? I do!

Get up now; grab a piece of paper and an ink pen. Make a list of who you want to call, who you want to convince, who you need to convince.

Time is wasting!

Chapter 14

United Citizen's Anti Katrina Plan

Can you believe it? It was April 2008, and I still had not gotten any answers to my enquiries concerning my "Social Security Disability Compensation!"

With 100% Service Connected PTSD; I can only handle so much disrespect before I have to blow up! See below:

Subj: ***Social Security***
Date: 04/01/2008 12:45:03 PM Taipei Standard Time
From: Congress.org [no-reply@congress.org]
To: Congress.org [no-reply@congress.org]

Thank you for using Congress.org Mail System

Message sent to the following recipients:
Secretary Paulson
Representative Mahoney

Senator Martinez
Senator Nelson
President
Message text follows:

[redacted],
Philippines, [redacted]

April 1, 2008
[recipient address was inserted here]
[recipient name was inserted here],

It has been my personnel experience that the Social Security Administration is run like a Communist Dictatorship. And the Orlando Florida Office of Senator Martinez, is a worthless accounting of the use of tax payer's money.

Further, I find it very difficult not to curse about it.

Thank you for your time!
Purple Heart [redacted].
Sincerely,
[redacted]

I still had the "Emergency Mass Migration Cities" on my mind during the first week of April. So, one day while re-reading it, I wanted to bring on board our country's two leaders.

I know, it can and will work!

With this in mind, I updated it with more of my thoughts! And e-mailed it to President Bush asking that it be forwarded to Sen. McCain by his office!

By doing this, and knowing that it was possible that Senator McCain had not had time to truly think it over, as he was very busy with the war, Osama bin Laden and running for the office of the President! I thought President Bush, Vice President Cheney and Sen. McCain could rally their resources and unite the citizens of the USA if they all 3 had a copy of the same idea. Hahaha, which I thought was a brilliant idea, and still do!

Also, I thought, because I had been the "Army of One!" who had given them the information leading to the capture of "Osama bin Laden," I would soon be receiving the reward money's! I certainly never thought that the reward money's would be given to the country of the Philippines, covertly, and saying that it was for there “World War II Veterans!” Wow that hurt! -----> Google it! www google com.

In fact, I thought my idea of "Emergency Mass Migration Cities," was genius!

See copy of that e-mail message, on next page:

Subj: ***Respectfully From Purple Heart*** [redacted]
Date: 4/3/2008 12:51:08 P.M. Taipei Standard Time
From: [redacted]
To: comments@whitehouse.gov
CC: vice_president@whitehouse.gov

Dear President Bush and Vice President Cheney.

I wish to thank you for your service to our country, our people and our way of life. I have always seen through the Demarcates ways of working against you and your hard work.

Truth is, you did win the war in Iraq and what has followed is, what would have happened here in the USA if we had not taken the war to Iraq, won it, and given "Osama Ben Laden Followers" a place to concentrate their efforts?

Thank you again,

Purple Heart [redacted]

Wonderful person, working for President Bush and Vice President Cheney! Please forward a copy of this email by fax or mail carrier to Senator McCain. It will "Unite the Citizens of the USA", create economical growth, and provide jobs and the increase in the value of the US Dollar.

Updated April 3, 2008

Kind and Understanding Sir! I do hope that you will be our next President. So I have taken this idea that has been floating inside my mind for a few years now and put it into an "email" for you to read. If it is needed, I can send a copy though US Mail!

It is not my intention to tell you what to do, but to pass the ball to you, so that you may be the quarterback, taking the ball and running with it...

It has to do with "Emergency Management Program's" and the possible benefits of uniting all Americans around a "Just Cause" called an "Emergency Mass Migration City."

If you find that this idea of a "Emergency Mass Migration City" could be useful to you and our people! Then please feel free to take this idea as your own, change it around or build on it. Put it before the people of our great country uniting them, and move forward with it as you feel might best, serve you and our people......

You will find that this plan (see below), as is, is short and not completely finished. It needs a "Think Tank" filled with minds that can conceive the more in-depth possibilities of an "Emergency Mass Migration" [Like what happened during "Hurricane Katrina"] and the benefits of an "Emergency Mass Migration City."

The purpose of sending you this plan is to save lives by helping "Elect Senator John McCain as the Next US President" by uniting the citizens of the United States of America in a [Just Cause] plan of action that will diminish the future effects of another "Katrina Type" catastrophe. Creating economical growth, providing jobs and the increase in the value of the US Dollar!

[Combined with a winning strategy on the "War on Terror", this can work, catapulting "Senator John McCain" into the "White House Oval Office" as "Commander And Chief!" A LION to lead our LIONS…]

Further, it is with great pride, as I look forward to that day in the future when I get to cast my vote in your favor.

Sir, I send this to you in happiness and with positive thoughts. The ball is in your court, take charge.

Sincerely,

Purple Heart [redacted]

United Citizen's Anti Katrina Plan

A "JUST CAUSE"!

United We Stand, For The People, By The People!

Author: Purple Heart - [redacted]

Sections of This Plan Include:

Terms:

Overviews of "United Citizen's Anti Katrina Plan."

Usefulness Of Plan

General Resources

Terms:

Just Cause: A justifiable cause that unites all citizens.

Anti Katrina Migration or Emergency Mass Migration: A major "US City" size, mass migration, brought on by immediate peril on a mass level. [Such as the result's brought about by the warnings put out as "Hurricane Katrina" approached and then the aftermath.

Even a dirty bomb could cause something like this!

An "Emergency Mass Migration", it has already happened and can happen again in our life time!]

Overviews of "United Citizen's Anti Katrina Plan"!

What is the "United Citizen's Anti Katrina Plan"?

A plan to build a "Emergency Mass Migration City."

A Plan that we can come together and agree on, standing united.

A plan that we can work together on!

A plan that can save lives!

A plan that can create jobs!

A plan "For the People by The People!".

We the people, "Of Our Great Country," can unite and prepare for an "Emergency Mass Migration" to a, pre-built "Emergency Mass Migration City."

Example Plans:

Plan A. A "United States Military Base" that is no longer needed as a Military Base and then turned into an "Emergency Mass Migration City." There-by, left empty until an "Anti Katrina Migration" or "Emergency Mass Migration" is implemented by the US President, FEMA and or Homeland Security.

Left empty until needed [only security and maintenance personnel on site 24/7 - 365].

Note: An entire city could migrate within 24 to 48 hours to a pre-built "Emergency Migration City." Opening Banks, ATM's, Schools, Shopping Centers, Hospitals, City Government Offices, Emergency Services, Airports, Law Enforcement Offices, Jails/Prisons, Animal Hospitals, with completely furnished housing ready for citizens to move into at a moment's notice! Even pastures for livestock…

(Not temporary type structures, but permanent structures that will last 100 years or more. [Naturally within a short time, citizens would begin emptying the "Emergency Mass Migration City" after their home city has been deemed fit to live in once again.] / Leaving us ready repeatedly for future mammoth size catastrophes and leaving a place for our "Grand Children" to go in an emergency, long after we are gone. (A "Green City" type place, so that our people can continue their day to day life's much sooner than from a temporary shelter...)

Even Wal-Marts, McDonald's, Corner Store, Mom and Pop, and Burger King type businesses could be up and running over night by a phone call to their suppliers with the address for their next inventory/supply shipment. [Managers could give the location (place of business), to employee's, to report to work at the next day, following the "Emergency

Mass Migration". The same could be done for schools "Teachers and Students"… City Gov and so on...]

Plan B. Build a "Emergency Mass Migration City". In an area that does not take up farm land or other natural resources. Like say, a desert! Complete with everything a city has, i.e. Baseball fields, parks, movie theaters, restaurants and so on, same as in the **notes** of **Plan A**, left empty until an "Anti Katrina Migration" or "Emergency Mass Migration" is implemented by the US President, FEMA and or Homeland Security. With only security and maintenance personnel, on site 24/7 - 365 days a year!

Plan C. Build 3 to 5 "Emergency Mass Migration City's", strategically positioned across the continental US. With another in Alaska! Left empty until an "Anti Katrina Migration" or "Emergency Mass Migration" is implemented by the US President, FEMA and or Homeland Security.

Plan D: Find 5 very small cities/towns inside the Continental US, like Berlin Ga. And build "Emergency Mass Migration City's" the size of the city named Valdosta Ga. around it. Leaving the "City Of Berlin" in-tacked, in the center, as the "Heart" of the "Emergency Mass Migration City"… This way, the 5 "Emergency Mass Migration City's" inside the "Continental US" could affectionately be called "The City of Heart's" location's 1, 2, 3, 4 and 5. Alternatively, "The City's With Hearts"… Even "Our City's With Hearts"! Alternatively, even, in which state it is put and then named after that state. Example: In Texas, "The Heart of Texas", In Florida, "The Heart Of Florida." and so on...

Plan E. Same as plan B, with sections for small cities.] Example: a "Emergency Mass Migration City" that is made up of sections that are ready for small towns or cities to move into in an emergency... Like floods or tornados that have left a small town or city completely destroyed. Giving the people a place where they can get back to a normal life over night, until their city is livable once again. And the rest of the "Emergency Mass Migration City" stands ready for any other "Emergency Mass Migration".] Further, the whole "Emergency Mass Migration City" could be built [laid out] like a pentagon, with each section for small cities. And as a whole, one large city!

[Setting the example for all countries to follow. i.e. Mexico and all of Central and South America and, EU, Russia and China and so forth.. Creating economical growth worldwide! **Note:** The UN could sponsor a number of "Emergency Cities" around the 3rd world, like say Africa or India that is not controlled by any government. However, by UN only....

Plus: If Saudi Arabia can build a complete city that is left empty, except for 2 weeks out of each year, then we can build many..]

Special Note: Our Homeless Veterans and other needy citizens could be given housing and jobs maintaining the empty "Emergency Mass Migration City's" on a temporary basis. Giving them the chance to hold their head up while they work toward a brighter future…

Example: A veteran could be given the job of operating a "Street Sweeper Machine" inside one of the "Emergency Mass Migration City's". Gaining experience and then receiving a VA loan to purchase a "Street Sweeper Machine" and obtaining a contract with a city government to clean their streets late at night.

Funding the "United Citizen's Anti Katrina Plan" Mandatory or Voluntary!

a. **Voluntary** - All employed citizens/non-citizens can and should have 1 to 5 dollars taken from their salaries by their employer's and sent in along with their Social Security taxes once a month on a voluntary basis. 10 million employed citizens/non-citizens sending in 2 dollars per month; would be 20 million dollars monthly toward the "Emergency Mass Migration City's".

b. **Mandatory** - All employed citizens/non-citizen's will have 1 dollar taking from their salaries by their employer's and sent in along with their Social Security tax's once a week on a mandatory basis with a voluntary 2 - 3 - 4 or 5 dollars added, and tax deductible... (The math looks good on this one also.)

c. **Mandatory or Voluntary** - Same as above but with a scenario like the "War Bonds of WWII ", where each contribution is repaid with interest like a US Savings Bond. It could be called something like, "The "United Citizen's Anti Katrina Plan" Savings Bond".

d. **Mandatory** - All Cities pay a yearly "Emergency City Tax" which covers every citizen and all businesses.

a. **Voluntary** - All "Emergency Management Program's" of the US Government fund the "Emergency Mass Migration City's".

Usefulness of Plan

"Insuring all construction material" is made in the "USA" and "all contracts" are with "small or family owned businesses", this would catapult economical growth, providing jobs and the increase in the value of the US Dollar.

Uniting the Citizens of the United States of America in a [Just Cause] plan of action, that will diminish the future effects of another "Katrina Type" catastrophe. By providing

"Emergency Mass Migration City's" that will be used on a 2 week, 2 month or even a 2 year basis only.

Note: There are certain skills to have when conducting mammoth size projects, such as building and maintaining "Emergency Mass Migration City's". It is best to have the US President, form a team of planners.

General Resources

"The People of the United States" is the greatest resource the world has ever known. United, they have undertaken "Mammoth Size Projects" in the past and can once again.

We the People, "Can and Will, Stand United In an Anti Katrina Plan".

Further Notes:

The US Military could use the surrounding areas of the "Emergency Mass Migration City's" as live mission training areas. Gaining mission experience in "Homeland Security" and protecting the "Emergency Mass Migration City's" at the same time.....

For whatever reason, it was never presented to the American people, until now, in this book! Please re-read it, consider it, and let's "Stand United!" Moving forward into the future, with "Emergency Mass Migration City's" sitting in wait, should they be needed?

We can change the world, let's get started!

I received an email asking me to cast my vote concerning water boarding, and I did! Then I received this automatic reply! See copy below: Don't you just hate it when you receive an automatic standard reply? I do!

Subj: ***RE: Your response from Senator Bill Nelson***
Date: 04/10/2008 05:00:44 AM Taipei Standard Time
From: Bill@billnelson.senate.gov
To: [redacted]

Please do not reply to this e-mail. If you need to send another message to Senator Nelson, please use the form on his Web site: http // billnelson senate gov / contact / index cfm # email.

Dear Mr. [redacted]:

Thank you for contacting me with your concern about the treatment of detainees and my position on the matter.

The methods our government uses to gather intelligence are a matter of serious concern to me, and a topic that I monitor very closely as a member of the Senate's Intelligence, Armed Services, and Foreign Relations Committees.

The Senate recently passed a bill reauthorizing Federal intelligence programs (S. 1538). In the wake of recent controversy over the use of a tactic known as water boarding, that bill included a provision requiring intelligence agencies to adopt the Army Field Manual's procedures for conducting interrogations. I supported passage of this legislation.

I appreciate your taking the time to share your views with me, and please do not hesitate to contact me again in the future.

P.S. From time to time, I compile electronic news briefs highlighting key issues and hot topics of particular importance to Floridians. If you'd like to receive these e-briefs, visit my Web site and sign up for them at http // billnelson senate gov / news / ebriefs cfm.

Then I received another suspicious email, see it next:

Subj: ***Fwd: From Iraq***
Date: 04/12/2008 08:12:30 PM Taipei Standard Time
From: [redacted]
To: comments@whitehouse.gov
CC: vice_president@whitehouse.gov

Wonderful person working for President Bush and Vice President Cheney! Would you forward this to the CIA and FBI please...

Thanks,
Purple Heart [redacted]
[redacted]

From: marde@fka.att.ne.jp
Sent: 4/12/2008 1:14:44 P.M. Taipei Standard Time
Subj: From Iraq

Good day to you,

My name is Sgt. Marc Krois "US Force in Iraq", I am serving in Iraq (Mosul) and I will need your assistance to take care of some financial matters for me and of course you will be adequately compensated.

I have to urgently evacuate the sum of $30,000,000.00 (Thirty million US dollars) clean legal notes and all in US$100 bills out of here to a safe place, having found a secure way to do so.

5 million US dollars (US$ 5,000,000) will be your share for rendering this assistance.

SOURCE OF MONEY:

Some money in various currencies were discovered concealed in barrels with piles of weapons and ammunitions at a location near one of Saddam's old palaces during a rescue operation several months ago and it was agreed by the few of us present that the money be shared amongst us, though this was an illegal thing to do, but I tell you what? No compensation can make up for the risks we are taking with our lives in this hellhole.

Click on this links to read about event that took place here http://news.bbc.co.uk/2/hi/middle_east/2988455.stm

The above figure was given to me as my share and for several reasons, concealing this volume of cash here has been a big problem/source of distraction to me. I have now found a temporary solution with the help of a lady who is working with an international organization here, She does not know the real contents of the packages used in concealing the funds and must not know rather believes that the packages belong to an Asian American who died in an air raid and before giving up, he trusted me to hand the packages as it is to his business associate in Europe.

One request I will like that you keep this information to yourself. Should you have any reasons to reject this offer, please destroy this mail, as any leakage will be too bad for us. I want to have a very peaceful retirement life after my service here and I rely so much on this fortune to provide that for me.

If this proposal is acceptable by you, I await your urgent e-mail so we can discuss more.

Semper Fi. / Sgt. Marc Krois / ***"US MARINES"***

I had sent a respectful email to Vice President Cheney, thanking him for his many days, years of hard work for our country, our people and our way of life. His office sent a reply; see below/next:

Correspondence from Vice President Cheney
From: Vice_Presidential_Correspondence@ovp.eop.gov
To: vice_president@whitehouse.gov
Sent: 4/11/2008 4:31:57 A.M. Taipei Standard Time
Subj: Thank you

Dear Friend:

On behalf of Vice President Cheney, thank you for your e-mail expressing support. He sincerely appreciates your thoughtfulness in taking the time to write and sends his warmest regards.

Jenny Rose Folsom
Special Assistant to the Vice President for Correspondence

*****

It was nearing the middle of April 2008, and I was getting nowhere, with everyone in the US Government that I had asked for help, concerning my "Social Security Disability Compensation."

In fact, the only Senator or Governor or even President Type individual, which had enough respect to reply about my encounter with "Osama bin Laden", was Senator McCain! So I contacted him (By putting it into the face of everyone!), about my many problems with the "Social Security Administration" and "Senator Mel Martinez."

I had already sent a message through Senator McCain's contact form on his website about, after receiving respect from his office, that I was not going forward with names and address that I had told Senator Martinez I was going to do by March 7th, in an earlier email!

Which was that, I would not send to the news channels "names and address" of the people responsible? Because I never want to cause "Senator McCain" any problems, and I still feel this heartfelt respect for "Senator McCain" even today! See copy of email below:

"Senator McCain" is held by his office, to keep all "TOP SECERT" information to himself, like my "SALUTE REPORT." Even if it would help him get elected as President by leaking it to the public! I am not, so let me give you more info for your investigation!

Subj: ***Re: Thank you***
Date: 04/22/2008 12:01:42 AM Taipei Standard Time
From: [redacted]
To: Vice_Presidential_Correspondence@ovp.eop.gov
CC: comments@whitehouse.gov

Hello, wonderful person working for President Bush and Vice President Cheney!

Would you please fax or mail a copy of this to Senator McCain, see below? And please tell Vice President Cheney, thank you for putting a smile on my face when he was joking around at that dinner the other night. Which was on CNN?

Message for Senator McCain follows:

Hello, wonderful person working for Senator McCain!

Do you remember me? I am the Purple Heart that sent Senator McCain the SALUTE REPORTS about "Osama Ben

Laden" here in the Philippines and the SALUTE REPORT about the security at SUBIC Bay!

I am being "Punished and Discriminated" against, by members of the Social Security Administration for 12 months now and, by the office workers at the Orlando Florida Office of Senator Mel Martinez…

Can Senator McCain help?

It hurts me to know that I have only been able to send $50.00 to Senator McCain's campaign when I know how badly he needs the funding…

Senator McCain, Sir! If I ever did anything that hurt your campaign for the "Office Of President, Of The United States Of America." I could never forgive myself! So please Sir, advice me! "What do I do concerning this "Punishment and Discrimination" against me (Purple Heart ****), a "Combat Wounded Veteran, of the 82nd Airborne, US Army?

I am the one that President Bush went on TV about in 2004, asking that all veterans with PTSD come back to the USA.

I had left because of many problems causing me stress. And my VA Doctor had advised me to get away from anything causing me stress. That led me to take my son to Panama where he was born while I was serving there in the US Army. I understand that you "Senator McCain", was also born there when your father was in the US Navy...

While there, I used a forwarding address in Miami for all my mail. And when the SSA sent forms for me to fill out concerning my "disability compensation," I filled them out and returned them like, I had done for over 10 years, letting them know that all my records were with the VA.

I closed that Miami address in June 2006. And returned to my home in [redacted] Florida. After returning to Florida, I went to the VA Outpatient Clinic in FT Myers Florida and found that they were using my father's medical records while prescribing medication for my heart. My father had died in 2005. I stop going there for that and other reasons.

On Feb 06, 2007, my son and I came here to the Philippines. During April 2007, the Social Security Administration cut off my disability compensation, because of a "FORM", they sent to my old Miami address. A "FORM" I did not receive and could not have sent back... Please see further below, a copy of a letter, I sent to Senator Mel Martinez and notice the dates...And know that it was only after I sent an email to President Bush that Senator Mel Martinez's office ever contacted me here in the Philippines by email. And please know that many times I have asked for an update on my case, and was given the run around for months. Then they stopped contacting me here in the Philippines and sent letters to my home address in Florida. And it was only after I called home that I found out what the letters said. Which was that "The SSA was sending me a FORM to fill out"?

It was Wed April 2, 2008 that I called the Social Security Administration Office at the US Embassy in Manila Philippines and asked for an update. I was told that the Medical and Mental examines by the VA I had on Feb 27, 2008 in Manila, was not good enough because the SSA is a different org. and that they wanted a different exam to be done by them. I said "When will it be?" and I was told that I would receive an appointment in the mail. I then asked where would the exam be, and was told "That it would be at the VA Outpatient Clinic in Manila". Which is where I had just gone for the same thing on Feb 27, 2008?

And to date (April 21, 2008), I have heard nothing further from them.......How long does it take? I have been waiting for 12 months...

This is a "Cloak And Dagger Strategy" for "Punishment and Discrimination" against me "Purple Heart [redacted]", a "Combat Wounded Veteran, of the 82nd Airborne, US Army. !.... And Senator Mel Martinez "Orlando Florida Office Personnel", are helping to put the knife into my back by the insulting way his "government employees" stop sending updates by email and then send them by letter, "info I need", to my Florida address. When they know full well, that I am here in the Philippines, sitting in my apartment 3 weeks out of each month, so that I can make what little money, I have last until next month. When history has proven, that the Philippine Government "Office Of Immigration" will find a way to get more and more money out of me..... And the Office of Governor Christ of Florida knows this is going on and does not help me.

Did you also notice that the US dollar to the Philippine peso started falling around the same time that my disability compensation was cut off? And that it did not start to stop until around the time I found out the truth as to the reason my disability compensation was cut off and began to fight back. Notice the dates below!

However, as history has proven, it is a good thing because I was able to send a SALUTE REPORT concerning "Osama Ben Laden" to you, President Bush and Vice President Cheney. And I wish you could somehow let me know how it turned out. Going by what I see on CNN and the President of Australia giving a "SALUTE" to President Bush after they talked. I assume it went well....

Not once, has anyone apologized for this "Punishment and Discrimination" against me (Purple Heart [redacted]), a "Combat Wounded Veteran, of the 82nd Airborne, US Army, by the SSA.

I am a US Veteran and this is not the highest quality health care, mental health care and or rehabilitative care in the world. Also, I am in the Philippines! This is a 3rd world country and the health care is just that, (3rd World).

Question: Is this a "Cloak and Dagger Strategy" by the Social Security Administration, to punish me for traveling? Are they going to do this again, if I go on a road trip to visit family members for a few weeks? And I am not home to receive their letter. Justifying the use of an ankle bracelet on any veteran that has "Service Connected PTSD"?

Do you know that I was following doctor's orders to get away from whatever is causing me stress... Just to end up getting a knife in the back by a mindless paper pusher sitting at a desk inside the office of their private little dictatorship? (SSA)

Please Sir, look up the form that they sent to me! The Form that caused all these problems… There you will plainly see, at the top of the front page, it says "If I Do Not Fill Out and Return The Form, I could Lose My Disability Compensation". It says nothing about the SSA having to make sure it made its way into my hands!

It has been 12 months now and the money that they have held back from me, adds up to over $10,000.00. Yes Sir, that hurts....

I'm sure that you understand, when I say that it hurts me deeply when I cannot send more than the $50.00, I sent last month as a contribution to your campaign.

Please see below, a copy of the letter, I sent to Senator Mel Martinez, and the email from john.mac@mail.house.gov that again proves that all of this is because of a form that was sent to me in Miami when I was here in the Philippines.

And please advise me as to what I should do. Because I stopped contacting the TV News in Florida after receiving the respectful email from your office. Where I was going to expose all information about those in government that were "Punishing and Discriminating" against me, (Purple Heart ****), a "Combat Wounded Veteran, of the 82nd Airborne, US Army…

Again, I do not wish to harm your campaign. However, I must fight in defense of myself, so that I recover my disability compensation....
Copy's follow.....

(Note added while writing this book: what you will read next, is a copy and paste out of a file, which I had written and printed out, then sent to the Orlando Florida Office of Senator Martinez by FEDEX. Because, at that time I could not send any email messages to him!)

From: [redacted]
[redacted] Philippines 2200

To: Office of U.S. Senator Mel Martinez
315 East Robinson Street
Landmark Center One, Suite 475
Orlando, Florida 32801

Dear Kind Understanding and Respected Senator Martinez. My name is [redacted]. Social Security number [redacted]

On the day of 6 Nov, 2007, I sent by FedEx, to your Orlando Office, a signed "Privacy Act Consent Form" along with other documents concerning my ongoing battle with the "Social Security Administration," about my disability compensation. Which had been turned off sometime around April 2007? The FedEx tracking number is [redacted]. I was told that it should arrive on the 8th of Nov 2007. As of today "Jan 8, 2008," I have not heard back from your office.

On 6 Dec 2007, I visited the VA Regional Office in the city of "Manila, Philippines." And found that the "Social Security Administration," has an office inside the same building. I went to their office and explained the problem I was having. Which is, that my "Social Security Disability Compensation," had been turned off, and that I had only found out about their letter to my old Miami Florida address, though a copy of their letter that was sent to me by email, in an attached file, from john.mac@mail.house.gov, see copy of email further in this letter)...?

I was then sent in, to talk to a Claims Examiner. And she told me that the whole problem had nothing to do with my continued disability. However, It was due to the fact that, I did not fill out and return a form that was sent to me by mail to my Miami address.

I then pulled out my passport and proved that I had been here in the Philippines sense Feb 6, 2007. And had never received the form, they had sent to my old Miami address. "Which was closed during June 2006, when I was moving to [redacted] Florida?" Where I lived until coming here to the Philippines for a visit... Furthermore, proving that I had not been in Miami during July 2007 when they had sent the letter that was later sent to me by john.mac@mail.house.gov, see copy of email further in this letter)..

She (The Social Security Disability Compensation Claims Examiner.) then handed me a Disability Report - Appeal - Form SSA-3441-BK. And said to take it home, fill it out and mail it back to her! Which I did; "Please, see a copy of that form (Filled out), that I am sending along with this letter to your Orlando office."

Update: Yesterday, Jan 7, 2008 - I called the Social Security Administration Office in Manila Philippines at 02-301-2000 extension-2621 and talked to Social Security Rep. "Gilbert Go." As he looked up my file on his computer and read it to me, I could tell that nothing has been done to resolve any issues. And I explained it all over again, just as I have been doing for months. He then explained that the "Claims Examiner" that I had talked to on 6 Dec, 2007 had "resigned." And that a new "Claims Examiner" would see my case. Further, he said that her name was "Patricio Everly," and to call her in 2 weeks at 02-301-2000 ext: 2459.

Dear Understanding Senator Martinez, I need this understood. I have been "permanently unemployable" sense 1991. And have depended on Social Security Disability

Compensation for all these years. I want a Federal Judge (even the Supreme Court) to hear my case as soon as possible. So that the Social Security Administration Office in Manila Philippines, "Claims Examiner" , "Patricio Everly," at 02-301-2000 ext: 2459, be demanded to turn my Social Security Disability Compensation back on and pay it retroactive back to the last month due me. And to stop this Social Security Administrative mistake, of turning off my Social Security Disability Compensation because of a form I never received. And never should have been sent to me because I have been "permanently unemployable" sense 1991.

Please have someone from your office contact me through email at [redacted].

*****

From: john.mac@mail.house.gov
To: [redacted]
Sent: 10/31/2007 5:00:30 A.M. Taipei Standard Time
Subj: RE: Your Social Security Issue (CONGRESSIONAL INQUIRY)

Mr. [redacted] after speaking to the Social Security Administration on your behalf, they forwarded information to me, which are contained in the attachment. Your benefits had been terminated by the Social Security Administration in February of this year. This was due to the fact that the Miami filed office had been trying to correspond with you to submit a new up- to- date medical evaluation. This evaluation is required by the Administration to substantiate your benefits. B/c you had failed to do this and had not responded to Social Security your benefits had been terminated.

I had also inquired with the Veterans Administration and they have a pending claim for a spouse on your record and that is currently in the review process. No determination has been made to date.

Senator McCain! Please advise me as to what I should do.
Purple Heart [redacted]

*****

Subj: ***Fwd: Priveledged to meet you.***
Date: 04/23/2008 08:20:21 AM Taipei Standard Time
From: [redacted]
To: comments@whitehouse.gov
CC: vice_president@whitehouse.gov

Hello, Wonderful person working for President Bush and Vice President Cheney. Do you know who might be able to get me off of the list that these people have...Please see forwarded copy below....? As you know, I have 100% service connected PTSD, and it seems like every scam artist in the world has my contact info.....?

Also, please tell President Bush, that it made me smile to see him wishing good luck to the Military Captain that was on the game show... I saw it on CNN....And it is good to see the President smiling......

Thank you,
Purple Heart [redacted]

****

From: nelson.botha@gmail.com
Reply-to: nnbotha@gmail.com
Sent: 4/23/2008 7:08:17 A.M. Taipei Standard Time
Subj: Privileged to meet you.

From: Mr. Nelson Botha

Dear Friend,

I am Mr. Nelson Botha an out of wedlock son of late Mr. Pieter Willem BOTHA. The Former Prime Minister and First State President of South Africa.

Please I seek your assistance for the investment of USD $12,000,000.00
{Twelve Million United States Dollar} which I inherited from my late father.

As you may know, my late father was the President of South Africa during the Apartheid-era and this has been a major setback for me since his demise in 2006. Though my late mother was a black South African, yet I am being treated like a stranger in my own country hence my decision to relocate to Europe or America as soon as I am able to lay my hands on my inheritance.

With immense pressure from The Old ANC leaders, the present Government of South Africa has sized all my father's properties in South Africa and frozen all his account with local banks here in South Africa, as my father is viewed by majority of locals as a monster, especially by middle-aged people, because of his role during the apartheid regime and refusal to release Mr. Nelson Mandela, whom he named me after.

Before his death, my father drew my attention to a sum of USD
$12,000,000.00 {Twelve Million United States Dollar} that he had stashed
away from me as an inheritance.

This fund is presently under the custody of a Security & Finance Company in UK as Family Valuables, fully documented under my name. I have the deposit papers in my possession.

Our family friend who is a lawyer and currently seeking asylum in UK, said it was my late father's instruction that I should look for a NON-SOUTH AFRICAN to claim this fund and not do so personally, because the fund was shipped out of South Africa as payment to a Foreign Contractor. He has further made arrangements with the Security Company to release the deposit to anybody that I nominate and the company has agreed to do that as long as a proper authorization letter from me is presented.

Hence, my purpose of contacting you to help retrieve this fund from the Security Company. I do not want this company to send the fund to South Africa, based on the fact that the South African Government will confiscate it, as they have done with other of his fund.

This is the only fund that I will have to survive on and I am not ready to lose it. The South African Government is not aware of this fund. It is secured and has a clean backup, with all depository documents in place.

I want you to stand as Family Partner to receive this fund and lodge it into your account for investment in your country or any part of Western World.

I wish to leave South Africa as soon as the fund goes into your account. And for your assistance in this noble manner, I am willing to offer you 20% of the total fund. Not only that, you shall also be the MD/CEO of whatever investment I make with my own share of the money, because I do not want my name to appear as a Director of any business abroad. This is in a bid to get the South African Government off my back forever.

I will come over to meet you in your country to invest this money with you as my business partner. We shall sign an Asset Management Agreement as soon as I meet with you in your country.

I will give you more details on the next step of actions as soon as I get your favourable response.

I expect to read from you as soon as possible. Please contact me on this email address only: nnbotha@gmail.com

Regards,
Mr. Nelson Botha

*****

Subj: ***Fwd: (no subject)***
Date: 04/24/2008 01:31:58 PM Taipei Standard Time
From: [redacted]
To: comments@whitehouse.gov
CC: vice_president@whitehouse.gov

Dear Sir, I have asked that somehow the FBI or CIA get me off of these peoples list of who to scam. To date I am still on the list. See copy below:

Please Sir, have someone insure that the FBI and CIA do their job.....

Thank you kind Sir;
Purple Heart [redacted]

*****

From: info@fedex.com
Reply-to: richard-raynor1963@hotmail.com
Sent: 4/24/2008 11:10:36 A.M. Taipei Standard Time
Subj: (no subject)

Customer Service.
Dear Customer!

We have been waiting for you to contact us for your Confirmable Package that is registered with us for shipping to your residential location. We had thought that your sender gave you our contact details.

It may interest you to know that a letter is also added to your package.

However, we cannot quote it's content to you via email for privacy reasons.

We understand that the content of your package itself is a Bank Draft worth of $800,000.00 USD. As you know, FedEx do not ship money in CASH or in CHEQUES but Bank Drafts are shippable. The package is registered with us for mailing by your colleague as claimed, and your colleague explained that he is from the United States but he is here in Nigeria for a three (3) months Surveying Project as he works with a construction firm in the Nigeria West Africa We are sending you this email because your package is been registered on a Special Order.

What you have to do now, is to contact our Dilivery Department for immediate dispatchment of your package to your residencial address. Note that as soon as our Delivery Team confirms your informations, it will take only one working day(24 hours)for your package to arrive it's designated destination.

For your information, the VAT & Shipping charges as well as Insurance fees have been paid by your colleague before your package was registered. Note that the payment that is made on the Insurance, Premium & Clearance Certificates, are to certify that the Bank Draft is not a Drug Affiliated Fund(DAF) neither is it funds to sponsor Terrorism in your country. This will help you avoid any form of query from the Monetary Authority of your country.

However, you will have to pay a sum of £100 GBP to the FedEx Delivery Department being full payment for the Security Keeping Fee of the FedEx Company as stated in our privacy terms & condition page. Also be informed that your colleague wished to pay for the Security Keeping Charges, but we do not accept such payment considering the facts that all items & packages that is registered with us have a time limitation and we cannot accept payment having known not when you will be picking up the package or even responding to us.

So we cannot take the risk to have accepted such payment in case of any possible demurrage.

Kindly note that your colleague did not leave us with any further information.

We hope that you respond to us as soon as possible because if you fail to respond until the expiry date of the foremost package, we may refer the package to the Nigeria Commission for Welfare as the package do not have a return address.

Kindly contact the delivery department (FedEx Delivery Post) with the details given below:

Contact Person: Richard Raynor
Email: richard-raynor1963@hotmail.com
Tel: +234 80-338-30603

Kindly complete the below form and send it to the email address given above. This is mandatory to reconfirm your Postal address and telephone numbers.

FULL NAMES:
TELEPHONE:
POSTAL ADDRESS:
CITY:
STATE:
COUNTRY:

Kindly complete the above form and summit it to the delivery manager on:richard-raynor1963@hotmail.com

As soon as your details are received, our delivery team will give you the neccessary payment procedure so that you can effect the payment for the Security Keeping Fees. As soon as they confirm your payment receipt of £100GBP, they will not hesitate to dispatch your package as well as the attahced letter to your residence. It usually takes 24 hours being an over night delivery service.

Note that we were not instructed to email you, but due to the high priority of your package we had to inform you as your sender did not leave us with his phone number because he stated that he just arrived Nigeria and he
hasn't fix any phone yet. We indeed personally sealed your Bank Draft and we found your mail contact in the attached letter as the recipient of the foremost package.

Ensure to contact the delivery department with the email address given above and ensure to fill the above form as well to enable a successful reconfirmation.

Do not reply this email because this email account is not monitored. Send your details to: richard-raynor1963@hotmail.com

Yours Faithfully,
Mrs. Margaret Blaire.
FedEx Online Team Management.
All rights reserved. © 1995-2008 FedEx.

This e-mail is only for the above addressees. It may contain confidential or privileged information. If you are not an addressee you must not copy, distribute, disclose or use any of the information in it or any attachments.

FEDEX
This message was sent using IMP, the Internet Messaging Program.

*****

Again, another respectful response from the office of Senator McCain. See copy on next page:

-----Original Message-----
From: senator@mccain.senate.gov
To: [redacted]
Sent: Tue, 29 Apr 2008 4:49 am
Subject: Correspondence from Senator McCain
DATE April 28, 2008

Mr. [redacted]
Dear Mr. [redacted]:

I wanted to take a moment to thank you for your recent correspondence. I appreciate your taking the time to assist me in keeping up to date. I found the material you referenced to be very informative.

I have passed along the information you forwarded to my foreign-relations and defense staff. I am confident that it will be useful. In the future, please do not hesitate to contact me on this or any other matter.

Again, thank you for the information. I certainly hope that you will continue to keep in contact on this and other matters.

Sincerely,
John McCain
United States Senator
JM/ab

Please Do Not Reply to this Message

On the night of May 13, 2008 Philippine time, I called home to Florida. While speaking to my mother, she informed me that my brother was there visiting her. I wanted to speak to him; I wanted to tell him the wonderful news. The news of finding, "Osama bin Laden!" The news about contacting "Senator McCain" with a copy of my "SALUTE REPORT" concerning my encounter with, "Osama bin Laden!" The news about the United States Navy ships being here and the fighter aircraft that was zooming through the sky! About the C-130 buzzing my home!

This was a wonderful time, and I wanted to share with my brother who had retired from the United States Navy; This wonderful experience of the United States Navy coming to the rescue, defending America, defending our way of life!

So I said to him over the phone, there's a lot going on here that I want to tell you about! But, I can't talk about it over the phone, is your e-mail address still good? He said, "YES!" So I went on to tell him that I would send him an e-mail and for him to e-mail me right back.

We went on to say our goodbyes as I felt enlightened by the thought that I was going to share with my brother, this truly patriotic experience I was having. We said our goodbyes and then I spoke to my mother again, then she and I said our goodbyes.

I went through the rest of the day waiting for my chance to write a long e-mail to my brother, about my encounter with "Osama bin Laden" and the days to follow.

The chance did not arise, but in the back of my mind, I knew that it would, and I would make this happen. I would share this experience with my brother through e-mail.

However, I have not, not even to this very day! Nor have I spoken about it to any family member, other than my

"Bubby" and my very good friend "Joe Alercia" at Lion Investigation Academy!"

After the phone call, as the hours went by, I rehearsed in my mind the words, the sentences that I would put in this e-mail to my brother, about my encounter with "Osama bin Laden."

The next morning while sitting at the edge of my chair while watching the cable news, the phone rang! It was my mother; she said "[redacted] has committed suicide!"

I paused, stunned!

She went forward talking, as my mind began to swim through spasms of bewilderment!

I said shocked, "He's gone?"

She said "Yes!"

I said once again in questioning disbelief, "He's gone?"

She said again, "Yes!" As I listened, she went on to tell me that my brother had received the call about his son the night before. He had driven alone, the 5 hour drive, back to his home, and his stricken family.

"My Heart was Broken!"

Disaster had engulfed my very soul!

I do not want to relive that here on this page. What I will do, is let you know! That it was the drug culture in America today, which made his life not worth living, in his own mind!

To compound this even further, it was the wife I spoke about at the beginning of this book. Who betrayed his young soul, and pulled him into a living hell?

As you read on through the days that followed, know in your heart, that the stress brought on from this knowledge, affected me deeply. To the extent of filling my mind with a desire to strike out at all things, I felt were wrong!

How could I be of any help? How could I roll back time? How could I find the words necessary to extinguish the flames of this living hell, my brother and his family were in?

I had not the money to fly home!

I cried rivers of thick sobbing tears, out loud and intently, with feelings so profound as not to let this moment, be ever taken lightly! I cried out decreeing with my voice energized with my very DNA:

"LUCIFERRRR, WHAT HAS'T THOU DONE…?"

This was my flesh! This was my blood! This was the son of my brother, who had so quickly, without second thought! Reached out his hand, and pulled to safety my little "Bubby." When, he was going under for the last time. While drowning in the dark waters of Fish Eating Creek Florida… As I watched in horror, screaming because I was stuck knee deep in the mud, on the creek bank, and could not get to "My Little Bubby" in time!

The nightmare, I had had back in the Gulf War! Where "My Little Bubby" was drowning, and then a voice said; "It's ok, your brother has saved him, was coming true!" "You see, in this case, my brother, and my brother's son are the same!"

To my brother: Sir, I am so sorry for your loss, and for your painful sorrow!

I sent a letter to the office of "Senator McCain" concerning this matter. A few weeks later, it was reported on the News, that the number of suicides in the military community had increased considerably!

"Thank you "Senator McCain" kind Sir, for your respect!"

Earlier in this book, I informed you of the colors that al-Qaeda and "Osama bin Laden" used to make an announcement covertly. Which was the same color of the prisoners at Qintunomo Bay Cuba?

Now read on, find the color orange, and who is wearing it. Also, know that they no longer wear it today, after my pointing it out, and the entire Philippine Military restructure!

This "One Man Army" has been changing the world!

Subj: ***Purple Heart*** [redacted]***reporting.. [Incident: 080302-000024]***
Date: 05/14/2008 04:10:11 AM Taipei Standard Time
From: U.S. Department of State [usdeptstate@mailnj.custhelp.com]
To: [redacted]

Recently you requested personal assistance from our on-line support center. Below is a summary of your request and our response.

Thank you for contacting the State Department.

Subject : Purple Heart [redacted] reporting…

Discussion Thread

Response (Support Agent) - 05/13/2008 04:10 PM
The Bureau of Public Affairs is in receipt of your message. Thank you for contacting the U.S. Department of State.

Question Reference #080302-000024

Date Created: 03/02/2008 10:16 AM
Last Updated: 05/13/2008 04:10 PM
Status: Solved

[---001:000631:50246---]
-----Original Message-----
From: noreply@contact.dhs.gov
To: [redacted]
Sent: Sun, 2 Mar 2008 11:03 pm
Subject: Contact Us Feedback
Thank you for your submission. The following information has been submitted.
User Name: [redacted].
User Email: [redacted]

To: securitythreats@dhs.gov
Category: Security Threats
Message: Purple Heart [redacted] reporting..

Situation: Very Poor security at Subic Bay Philippines for US Navy Ships:

How I know!

Today my son [redacted] and I were looking for a street with no traffic, so that I could give him a few lessons on how to drive a stick shift (5 speed Toyota). We left the house at [redacted], [redacted], [redacted] city, Philippines and drove to Subic Bay. After driving though a few streets full of traffic, I turned toward the Bay, came across a security road block and turned to the left to avoid it. As I kept driving on, we came to a road with a security check. So I turned right to avoid it. Driving on southward I came to a closed gate and a turn to the right. So I turned right and drove west. Which brought us to a stop sign...? I stopped and looked right (North) and seen a road block by Subic Security, so I turned left to avoid it.

Bang........... I'm looking straight at an American Navy Aircraft Carrier, American soldiers are in there street cloths walking away from the ship and Philippine soldiers are dressed in their New BDU with the **orange** part of their camo uniforms showing brightly in the midday sun.

I turn right, west; at the corner and drove along the street not believing how easy it was to accidentally drive though all the security road blocks. And right beside within a few feet of the soldiers.

As we drove on westward we came to a security road block again, so I pulled into a parking spot in front of the ship named "Niagara Falls". As I looked at the security guards at the road block, I could tell that one of them had spotted me, but was not sure what to do.

Size: 7
Activity: Not providing secure security.
Location: Subic Bay Philippines
Uniform: Subic Bay Security Uniforms
Time: 2:00 pm Philippine time, 2 March 2008
Equipment: 2 trucks, 1 gate, 12 orange traffic cones.
Copy sent to: comments@whitehouse.gov
vice_president@whitehouse.gov and CIA though there website contact form…..

Date and Time of message: 3/2/2008 10:03:09 AM

*****

Note to Reader: If you've ever been in the military, you might already know about salute reporting. If not, on the Internet, go to http // www geocities com / esm2nddiv / salute html.

There you will find the following information about salute reporting. Here you will see that the letter S. stands for the word S I Z E. In my salute report, you see that the S sometimes stands for SITUATION. This is because while I was in the Army, we used the S in SALUTE for SIZE or

SITUATION. And attempted to put the Who, What, When, Where, How, here. You may want to put the Who, What, When, Where, How, inside the A for ACTIVITIES section, while writing your salute report.

You will also see that I use the U for uniform, were below, the U is for unit. Either way will work just fine! Just make sure to put as much to memory as you can, then use the "S.A.L.U.T.E REPORT" format to help you put it into context for the reader…

S.A.L.U.T.E REPORT (Use this format when reporting to your unit commander) http // www geocities com / esm2nddiv / salute html

The **S.A.L.U.T.E**. Report / When reporting any intelligence information to Headquarters or other Units the S.A.L.U.T.E. format should always be used. This will ensure all information is accurate and complete.

S ize A citivity L ocation U nit T ime E quipment.

S=SIZE of enemy force. Actual numbers if possible. (Squad=12, Platoon=38, Company=145, Battalion=600)
Details:

A=ACTIVITIES of the enemy. (If moving, give direction and method of transportation. Take your time. Be specific. Include small details, even if they seem insignificant.)
Details:

L=LOCATION. 6 digit coordinates, or clear description.
Details:

U=UNIT IDENTIFICATION. (Markings, symbols, uniforms, vehicles, agency markings. Don't give up until you are sure about this one...)
Details:

T=TIME and DATE of sighting. Give Military 24hr time or am/pm.

Details:

E=EQUIPMENT carried by the enemy. (To include: type weaponry, webgear, electronics, night vision, body armor, vehicles, tents, etc.)

Details:

Additional information or photos attached: Also, put your 'Signature" here.

*****

I was getting concerned that "Senator McCain's" campaign was going through a downturn, and hope to help by somehow corresponding with Sen. Lieberman, who I was sure knew about the "SALUTE REPORT" concerning "Osama bin Laden." I received a reply from his office and was insulted. Simply, because it implied that my message had not been read, and that this message was a pre-determined automatic response. See copy of reply message on next page:

-----Original Message-----
From: correspondence-email@lieberman.senate.gov
To: [redacted]
Sent: Mon, 30 Jun 2008 11:24 pm
Subject: Correspondence From Senator Lieberman

TIME \@ "MMMM d, yyyy" June 30, 2008

Mr. [redacted]
[redacted],

Dear Mr. [redacted]:

Thank you for taking the time to contact me. I appreciate the opportunity to know your opinions on the pressing issues facing our nation. Regrettably, due to the huge volume of mail that I receive, I am only able to research and address comments sent to me from Connecticut residents. If you are not from Connecticut, you may want to consider sending a message to the Senators from your state of residence. You can do so by visiting http // senate gov / for a link to the web sites of each member of the United State Senate. If you are currently residing out of state, but are still a Connecticut resident or have a connection to Connecticut, please be certain to use your Connecticut address or indicate your Connecticut connection in the first paragraph of your email.

I value having the benefit of your thoughtful concerns, since I do receive detailed weekly reports from my staff providing a sampling of comments from across the country on timely issues before Congress.

Thank you again for sharing your views and concerns with me. I hope you will continue to visit my website at http // lieberman senate gov, for updated news about my work on behalf of Connecticut and the nation. Please contact me if you have any additional questions or comments about our work in Congress.

Sincerely,
A Joseph I. Lieberman
UNITED STATES SENATOR

JIL:gte

Chapter 15

International Incidence!

I waited until this chapter to write about my doctor's appointment at the VA Clinic in Manila during Feb 27, 2008. Because the SALUTE REPORT concerning it was deleted from my email account by person or persons unknown! Also, putting it all in one chapter will help you to see it in order of events! Your eyes will open widely when you read what happened when I went to my appointment at the same VA Clinic in Manila during May 7, 2008 and June 26, 2008!

Even so, first, take a look at this email below, next! Question: Was my email address given out to the US Embassy in Iraq? I had deleted the first email I received from them, thanking it was a scam. Then when I received this one a week later, I became very concerned!

Subj: **Please Forward To CIA**
Date: 06/23/2008 03:13:11 PM Taipei Standard Time
From: [redacted]
To: comments@whitehouse.gov
CC: vice_president@whitehouse.gov

Please Forward To CIA.
Question: Am I supposed to know this person? See below!
Geee, wouldn't I like to get involved!
Let me know if I can help.....
Thanks, Purple Heart [redacted]

From: alcom@masrawy.com
To: [redacted]
Sent: 6/23/2008 12:33:50 P.M. Taipei Standard Time
Subj: Request For Quote.

Attention: The General Manager,
We have sent you an earlier email after confirmation of your company from Embassy and we are yet to receive your response. We are requesting an answer to know if your company can quote for an urgent Joint Contracting Command contract to supply to both Iraq and Afghanistan. If yes, please do kindly send immediately your company's profile / products for immediate endorsement. Please do kindly treat very urgent and respond to this email address only.

Thank you,
Mr. Ali Al-Jawad Mehsin.
Iraqi Chancery,
Tehran Office.
Tel/Fax: 23982501.

Next, are copies of the automatic responses, after I had sent this email, by forwarding it to President Bush, and Vice President Cheney;

Subj: **(No Subject)**
Date: 06/23/2008 03:13:17 PM Taipei Standard Time
From: comments@whitehouse.gov
To: [redacted]

On behalf of President Bush, thank you for your correspondence.

We appreciate hearing your views and welcome your suggestions.

Due to the large volume of e-mail received, the White House cannot respond to every message.

Thank you again for taking the time to write.

Subj: **(No Subject)**
Date: 06/23/2008 03:13:16 PM Taipei Standard Time
From: vice_president@whitehouse.gov
To: [redacted]

Thank you for e-mailing Vice President Cheney. Your comments, suggestions and concerns are important to him.

Unfortunately, because of the large volume of e-mail received, the Vice President cannot personally respond to each message. However, members of the Vice President's staff consider and report citizen ideas and concerns.

Please visit the White House web site for the most up-to-date information on Presidential initiatives, current events, and topics of interest to you.

Thank you again for taking the time to write.

Reader; it is painfully obvious, that I need a private security agency of my own! This One Man Army needs to sleep, at some time during each day!

Below next, just like so many Americans, I expected to be a part of the stimulus plan. I have never received any Economic Stimulus Refund to date! It looks like I was left out!

Subj: **Fwd: Get 2008 Economic Stimulus Refund ($1800)**
Date: 06/24/2008 11:21:52 PM Taipei Standard Time
From: [redacted]
To: comments@whitehouse.gov
CC: vice_president@whitehouse.gov

Please forward to FBI.....it took me to some other website....

Thank you,
Purple Heart [redacted]

From: service@irs.gov
Sent: 6/24/2008 9:53:06 P.M. Taipei Standard Time
Subj: Get 2008 Economic Stimulus Refund ($1800)

Over 130 million Americans will receive refunds as part of President Bush program to jumpstart the economy.

Our records indicate that you are qualified to receive the 2008 Economic Stimulus Refund.

The fastest and easiest way to receive your refund is by direct deposit to your checking/savings account.

Please click on the link and fill out the form and submit before June 25th, 2008 to ensure that your refund will be processed as soon as possible.

Submitting your form on June 25th, 2008 or later means that your refund will be delayed due to the volume of requests we anticipate for the Economic Stimulus Refund.

To access **Economic Stimulus Refund**, please **click here.**

© Copyright 2008, Internal Revenue Service U.S.A. All rights reserved.

Subj: **(No Subject)**
Date: 06/24/2008 11:22:02 PM Taipei Standard Time
From: comments@whitehouse.gov
To: [redacted]

On behalf of President Bush, thank you for your correspondence.

We appreciate hearing your views and welcome your suggestions.

Due to the large volume of e-mail received, the White House cannot respond to every message.

Thank you again for taking the time to write.

*****

Subj: **(No Subject)**
Date: 06/24/2008 11:22:02 PM Taipei Standard Time
From: vice_president@whitehouse.gov

To: [redacted]

Thank you for e-mailing Vice President Cheney. Your comments, suggestions and concerns are important to him.

Unfortunately, because of the large volume of e-mail received, the Vice President cannot personally respond to each message. However, members of the Vice President's staff consider and report citizen ideas and concerns.

Please visit the White House web site for the most up-to-date information on Presidential initiatives, current events, and topics of interest to you.

Thank you again for taking the time to write.

Reader: I had been given an appointment at the Veterans Administration on February 27, 2008. After my encounter with Osama bin Laden, I had attempted to change my appearance by not shaving, and not getting a haircut!

In the letter, I had received informing me of the appointment, I was told to arrive one hour early. My appointment was at 9:00 AM. I arrived at 7:55 AM with my wife riding in the car along with me!

Once I had signed in, I walked outside to sit with everyone else that was in the smoking area. A few minutes later, I walked into the restroom next to the smoking area; it had not been cleaned for a week. The smell of urine was so strong that I had to hold my breath, finish what I had to do, and run outside into the fresh air.

An hour and a half later I was sitting inside the office of the veteran's psychologist on duty for my PTSD. Being under as much stress as I had been under since I walking up on "Osama bin Laden" and had been told by the man in black uniform, “Follow you die!” To include the other instances where I was followed, I was ready to talk to somebody that was paid to sit still and listen!

Immediately, I brought up the fact that my DD 214 had been in the Veterans Administration system since 1991. If anyone in the Veterans Administration was supposed to see it, then it was available on their computer. If not, then it was none of their business, and is why I had blackened out all information on my DD 214, when I turned it in at the front desk.

I then went into talking about being tortured with a "Dead Heart" for 3 1/2 hours, explaining that this was the beginning of my problems. She seemed to take notice of that and started typing into her computer.

After 50 minutes of my talking, she was ready to prescribe medication that would help me sleep. I stated that I did not want any medication, and that I had not come there for medication. What I needed was psychological tools, like, I had received in the past that had helped me greatly during times of extreme stress.

I was given a return appointment for May 7, 2008 and sent home! On the way home, it had taken us an hour and a half to get through the city of Manila and over onto the four Lane superhighway. Halfway down the super highway to our turn off, we stopped at a rest area/gas station.

As my wife walked to the public restroom, I went inside the convenience store, and stood in line. Twice someone came in the door behind me and cut in front of me in the line. This type of thing happens often in the Philippines as most people here, are very rude. Then a woman of around 26 years of age came in and just stood within 3 feet of me keeping me in her peripheral vision. This stood out as unusual, because she kept a stone face. I would see her many times again, until I reported it though a SALUTE report to Senator McCain. Then, and only then, did she and her team, stop following me... I'm sure another team has or will be assigned to follow me.... Too bad, they did not study under Joe Alercia of "Lion Investigation Academy!" If they had, they would have learned the error of their ways...

Once outside of the convenience store, I met with my wife, and we walked toward a snack bar. Just up ahead of us sitting at outside tables was an American man in his early 50s and a Filipino man in his late 20s. They were sitting facing each other.

As we walked by, I watched them in my peripheral vision and noticed that the American man had an expression of animosity as he watched us approach. The Filipino man, I noticed as we had passed them then turned right to go inside the snack bar, had a look of indifference.

I suspected that they were either from the US Embassy or were CIA and had been following us. So I looked to see if I noticed any vehicles in the parking lot that I had remembered from earlier in the day. Two matching style pickup trucks were parked next to each other, not far from where the two men were sitting.

In my mind, this confirmed my suspicion that the US government was having me followed!

Once home and situations allowed, I sent a SALUTE REPORT to President Bush, Vice President Cheney and Sen. McCain concerning my visit to the Veterans Administration outpatient clinic in Manila. Mentioning, the horrific odor and un-cleanliness inside the restroom!

Now it was May 7, 2008, I was back at the Veterans Administration outpatient clinic in Manila and took notice of the spit shined cleanliness of the outside restroom I had reported about. To include the firing of the ambassador to the United States Veterans Administration in the city of Manila, Philippines, and the appointment of a new ambassador!

What stood out the most was the security detail that was assigned to the ambassador. They were sitting around watching TV while she was in a closed-door meeting.

This was very unprofessional; I thought!

When she walked out of the office, they sprang to life and acted like professionals! On the way home we passed them head-on to see their emergency lights flashing, horns blowing as they forced traffic to the side of the street. You would think it was a convoy with the President of the United States.

Once inside the counselor's office, she wanted to prescribe medication to help me sleep. I complained that I did not want any medications that put me to sleep, I needed to stay awake.

She began to get bored to the extent she was almost falling asleep herself, as I was talking. So I spoke up and said, "Before I go, I need to get off of my chest something that happened to me in December. You cannot put it on your computer! This is, off the record!

As her eyelids nearly closed from her in and out of sleepiness, I began telling what happened to me in December at the office of immigration and my encounter with Osama bin Laden!

Maybe three sentences after I said the name "Osama bin Laden," her eyes open widely, and she said, "The coop!"
I replied "Yes!"

She immediately turned to her computer and started typing!

This set me on fire; I spoke up and said, "You cannot put that in your computer! Don't put that into your computer, this is, off the record. You are not supposed to put that in your computer, you need to stop!"

She turned to me and said "It's okay; I forgot something!", then continued typing as if she were an undercover operative sending an emergency message!"

I clamped my mouth shut and didn't say another word!

A minute later, she asked "Is there everything elsc?" I replied with "No!" She then rolled her chair back and said "Well, I have work to do, you can go now!"

As I stepped outside the door of her office and closed it, I turned right to go to the front desk. Immediately, I heard over the loud speaker intercom, her, call to another doctor, asking for him to come to her office.

I had been betrayed!

As soon as I returned home, I sent in a SALUTE report to President Bush, Vice President Cheney, and Sen. McCain concerning this betrayal!

A few days later, a hacker, got into my computer and destroyed many of the files, with my notes. Also, my email account, was nearly, totally erased of all corresponds to the U.S. Government. Then when I sent a complaint to President Bush and Senator McCain, **some were put back!**

Find the copy, of the reply from the office of Senator McCain on the next page:

-----Original Message-----
From: senator@mccain.senate.gov
To: [redacted]
Sent: Wed, 4 Jun 2008 3:36 am
Subject: Correspondence from Senator McCain

DATE June 3, 2008

Mr. [redacted]
Dear Mr. [redacted]

I wanted to take a moment to thank you for your recent correspondence. I appreciate your taking the time to assist me in keeping up to date. I found the material you referenced to be very informative.

I have passed along the information you forwarded to my foreign-relations and defense staff. I am confident that it will be useful. In the future, please do not hesitate to contact me on this or any other matter.

Again, thank you for the information. I certainly hope that you will continue to keep in contact on this and other matters.

Sincerely,
John McCain
United States Senator
JM/ab

Please Do Not Reply to this Message

I soon received a letter in the mail from Social Security saying that I was to report to the United States Veterans Administration outpatient clinic in Manila for an examination as to whether or not I was to continue receiving social security administration disability compensation.

On June 26, 2008 I arrived for my Social Security examination at the United States Veterans Administration outpatient clinic in Manila.

Your eyes will open widely! Because, the man that does the interview is the same Philippine Officer... I sent in a SALUTE report on, that while standing next to a military truck and other military men in uniform, during Jan 2008, who said, "There he is! Let's go, get in, let's go!"

They, with him in command, were assigned to follow me!

And now, he is doing the interview for the Social Security Administration of the U.S.A... Just how far, or how deep, does this type of spying go!

Allowing any non-U.S. citizen, a job, where they decide whether or not a U.S. citizen, receives or continues to receive disability compensation, is a betrayal against all Americans, and needs to be stopped.

Not only that, allowing non-U.S. citizens to work inside our embassy's, allows spy's a free rain through-out our in-tier government system.

When will we American's wake up?

What you are about to read, is the SALUTE report, I sent, concerning this examination!

Subj: **Prepare For An International Incidence!**
Date: 06/29/2008 10:23:19 AM Taipei Standard Time
From: [redacted]
To: comments@whitehouse.gov
CC: Vice_Presidential_Correspondence@ovp.eop.gov, vice_president@whitehouse.gov

Please forward to Senator McCain! Salute Report By Purple Heart [redacted]

Situation: Prepare For An International Incidence Reportable On CNN!

Activity: Because I was not home in Florida to receive the form to update my records! My Social Security Disability Compensation was turned off after I came to the Philippines on Feb 6, 2007. I had been receiving Disability Compensation from the Social Security Administration sense 1992. I have sense been trying to regain my Social Security Disability Compensation through the Social Security Office in the US Embassy, Manila Philippines that is manned by Non-US Citizens, also President Bush (comments@whitehouse.gov), Vice President Cheney (vice_president@whitehouse.gov), Governor Crist of Florida (Charlie.Crist@eog.myflorida.com), Congressman Martinez of Florida (Lane_Mahood@martinez.senate.gov), Senator McCain (http // mccain senate gov / public / index cfm?FuseAction = Contact ContactForm) also concerning my encounter with "Osama bin Laden," asking that it be forwarded to the CIA. This email has been removed from my email account without my permission from persons unknown.

Further: I was sent a notice of a doctor's appointment for 26 June, 2008 by mail. When I reported for this appointment, 1 Filipino man around 30 to 35, military officer style hair cut (hair, light brown in color), brown eye's, 5'5" to 5'6" in height, conducting a closed door forced interrogation. Never said his name!

I was forced to answer the following questions!

What is your Social Security Number?

Why did you come to the Philippines?

Why did you get married? How did you meet her?

What branch of the US military did you serve in?

What "Unit" were you in? Where is that "Unit" located?

From, what year to what year!

Did you serve in any wars?

What wars? From what date to what date?

What military schools did you complete? Where were these schools located?

What unit did you serve in?

Why did you leave the Military? How did you leave the military?

What jobs have you held?

Where?

What did you do while at that job?

Why did you leave that job?

Who controls your money?

At one point! He said: "I heard you talk a lot!"

Note: It is not the US Veterans Administration. It is Philippine Doctor's posing as the US Veterans Administration gathering information right under the nose of the US Ambassador while she wears a vest that has writing on it. It is light brown in color and has the words "Project Commander" embroidered on the back. As she walks around and points, a crowd of 5 or 6 Filipino men and women follow and look at what she point's at.

Note: I was very angry about this questioning and voiced my objection loudly! Each time I objected to the question, his hand came up to my face, palm out. He then tried to manipulate me into voluntarily going into a locked and guarded hospital ward, without a lawyer, US rep or family member! (The man told the office clerk to send me to the "psychiatric ward" of the hospital. A locked ward! I am to leave my rental car at the VA clinic. Take a taxi to the hospital and walk into the locked ward without a lawyer or US rep.

Note: Each question answer's the area of my DD214, which I had blackened out with a magic marker, when I gave them my DD214 months ago.

Note: I fully intend not to cooperate any further. I lied and said I was leaving the Philippines on July 1, 2008…..In truth, until my social security disability compensation is turned back on and paid retroactive back to the day it was turned off, I do not have the money to get my son and I back to the US together. I will not abandon him here.

Location: Manila Philippines, VA Outpatient Clinic Pasay City.

Uniform: He was wearing: Faded Blue Jeans, Faded Blue T-Shirt, Walking shoes, White knee length button up medical coat, opened in front (not buttoned).

Time: 9:10 am until 10:11am, 26 June 2008, Philippine time zone.

Equipment: 1 office chair, 1 office desk, 1 office computer, 1 list of questions. 2 armed uniformed guards, outside front door.

More follows: not enough time to write it all now. I will send further info soon…..
Special note: I fully intend not to cooperate any further with any Non-US Citizen.
Signed: Purple Heart [redacted]

Copies prepared for: President Bush (comments@whitehouse.gov), Vice President Cheney (comments@whitehouse.gov), Vice President Cheney (vice_president@whitehouse.gov), and Senator McCain (http // mccain senate gov / public / index cfm?FuseAction = Contact ContactForm), CNN at news tips http // edition cnn com / feedback / forms / form11b html

*****

Subj: **(No Subject)**
Date: 06/29/2008 10:23:28 AM Taipei Standard Time
From: comments@whitehouse.gov
To: [redacted]

On behalf of President Bush, thank you for your correspondence.
We appreciate hearing your views and welcome your suggestions.
Due to the large volume of e-mail received, the White House cannot respond to every message.

Thank you again for taking the time to write.

Subj: **(No Subject)**
Date: 06/29/2008 10:23:28 AM Taipei Standard Time
From: vice_president@whitehouse.gov
To: [redacted]

Thank you for e-mailing Vice President Cheney. Your comments, suggestions and concerns are important to him. Unfortunately, because of the large volume of e-mail received, the Vice President cannot personally respond to each message. However, members of the Vice President's staff consider and report citizen ideas and concerns.
Please visit the White House web site for the most up-to-date information on Presidential initiatives, current events, and topics of interest to you.
Thank you again for taking the time to write.

*****

Find on the next page, the reply again, from the office of Senator McCain:

-----Original Message-----
From: senator@mccain.senate.gov
To: [redacted]
Sent: Tue, 1 Jul 2008 6:02 am
Subject: Correspondence from Senator McCain

DATE June 30, 2008

Mr. [redacted]
Dear Mr. [redacted]:

I wanted to take a moment to thank you for your recent correspondence. I appreciate your taking the time to assist me in keeping up to date. I found the material you referenced to be very informative.

I have passed along the information you forwarded to my foreign-relations and defense staff. I am confident that it will be useful. In the future, please do not hesitate to contact me on this or any other matter.

Again, thank you for the information. I certainly hope that you will continue to keep in contact on this and other matters.

Sincerely,
John McCain
United States Senator
JM/ab

Please Do Not Reply to this Message

READER: As you have read, from time to time, throughout this book. I have been betrayed many times through-out my life. Now, my life, had lead me to the Philippines and Osama bin Laden.

I had been going behind my wife, and reporting it to the U.S. President, Vice President, Senator McCain, the FBI and CIA. All, with the knowledge, that I was putting my neck on the chopping block. While, sitting back and watching the world change around me… Which had come about concerning, and after my SALUTE reports?

Now think for a moment, then ask yourself, "Does this man deserve more respect, then he is getting?"

The suspicious e-mails continued and I continued to report them, see next.

Note: They knew my full name!

Subj: **From Purple Heart** [redacted]

Date: 07/01/2008 06:27:12 AM Taipei Standard Time
From: [redacted]
To: comments@whitehouse.gov
CC: vice_president@whitehouse.gov

Please forward to CIA. Their contact form on their website, will not allow me to drop and drag a copy of this or any email.

Doing so, might help them find out who leaked my email address after I sent in the SALUTE report......

From: maudley@rocketmail.com
Reply-to: audley.maxwell@ymail.com

To: [redacted]
Sent: 7/1/2008 12:37:28 A.M. Taipei Standard Time
Subj: [redacted]<----*my full name!!!!*

From: The Desk of Audit Department,
Money Gram, Sg-Ssb Bank Ltd,
Faanofa Branch, Accra Ghana.

Dear [redacted], <----*my full name!!!!*

From the office of Audit Department, Money gram Sg-Ssb Bank Ltd branch.
Dr. Audley Maxwell is the Money gram new appointed Chief Auditor in charge of the Head office and controller to all Money gram Sg-Ssb Bank Ltd branch Offices in West Africa.
I want to inform you about some Funds that are in our possession that is registered in your surname and need to be transferred to you from our office here in Accra Ghana.
For further information i will advise you to contact me via my private email as stated below concerning the transfer process ASAP for a prompt remittance of the
funds to your nominated Bank account.

Name: Dr.Audley Maxwell
Email:audley.maxwell@ymail.com
With kind regards i look forward to hearing from you soon.

Best Regards, Dr. Audley Maxwell.

Subj: **(No Subject)**
Date: 07/01/2008 06:27:19 AM Taipei Standard Time
From: comments@whitehouse.gov
To: [redacted]

On behalf of President Bush, thank you for your correspondence.
We appreciate hearing your views and welcome your suggestions.
Due to the large volume of e-mail received, the White House cannot respond to every message.

Thank you again for taking the time to write.

*****

Subj: **(No Subject)**
Date: 07/01/2008 06:27:19 AM Taipei Standard Time
From: vice_president@whitehouse.gov
To: [redacted]

Thank you for e-mailing Vice President Cheney. Your comments, suggestions and concerns are important to him Unfortunately, because of the large volume of e-mail received, the Vice President cannot personally respond to each message. However, members of the Vice President's staff consider and report citizen ideas and concerns.
Please visit the White House web site for the most up-to-date information on Presidential initiatives, current events, and topics of interest to you.
Thank you again for taking the time to write.

Below and to follow are my notes concerning that Day, of which I used while writing the Salute Report!

Notes:

Prepare For An International Incidence Reportable On CNN!

Prepare For An International Incidence!

Got out of bed at 4:00am
Left at 5:00am
Arrived at 8:15am
Doctor called my name at 9:10am

I had seen him before: he is the same Philippine military officer that was watching me when I was walking home from the city park, and said to his men: There he is, get in, let's go, let's go....

He said:

I heard you talk a lot!

He kept putting his hand up in my face as if to say "Shut up"!

Each time he asked a question, I answered. Each time I gave an answer, he asked me to explain. Each time I started to explain, he would put his hand up to stop me.

Many times he would say, stop, I don't want to hear it.

He read questions from a list.

He asked for my SS#. When he asked, I said, they already know my SS#. He then said, "Stop, I don't want to hear it, I have a list of questions to ask and that is one of the questions. (The SSA makes an appointment for me with the VA and they don't know my SS#. When in history has a non US citizen had the right to know a SS# of a US citizen.)

It felt like an interrogation.

I was forced to answer the following questions or face being thrown out on the street! To include a confrontation with the security guards; which could get me arrested!

What is your Social Security Number?
Why did you come to the Philippines?
Why did you get married?
What branch of the US military did you serve in?
What "Unit" were you in? Where is that "Unit" located?
From what year to what year!
Did you serve in any wars?
What wars?
What military schools did you complete?
What unit did you serve in?
Why did you leave the Military? How did you leave the military?
What jobs have you held?
Where?
What did you do while at that job?
Why did you leave that job?
Who controls your money?

Most of these questions, answer the information that I blackened out, when I turned in my D.O.D. discharge papers at the front desk. This information is in the system with the VA, and is available to anyone that has security clearance....

In front of me is a wooden frame with the name of the Doctor, I reported on a few weeks ago by the name of Eleanor Ronquillo. I asked" Is Doctor Ronquillo here today?" He said "No, maybe later!" Then looked at me through the corner of his eye! When I gave no reaction! He went on typing.

It is not the US Veterans Administration. It is Philippine Doctor's posing as the US Veterans Administration gathering information right under the nose of the US Ambassador while she wears a vest that has writing on it. It is light brown in color and has the words "Project Commander" embroidered on the back. As she walks around and points, a crowd of 5 or 6 Filipino men and women follow and look at what she point's at.

Someone went into my AOL email account and deleted the emails; I sent copies of to President Bush and Vice President Cheney. Then once I complained, they put some of them back......

The man told the office clerk to make me an appointment at the "psychiatric ward" of the hospital. A locked ward! I am to leave my rental car at the VA clinic. Take a taxi to the hospital and walk into the locked ward without a lawyer or US rep.

She called and said I was to report to the ward on the 8th. I lied and said I was leaving the Philippines on the 1st of July to return home to Florida. She then said that she would have to report that to the administration. I said "OK"! The call ended.

READER: On July 13, 2008! My Social Security Disability Compensation was turned back on, and paid retroactive, back over one year!

SITUATION RESOLVED!

I had an appointment sent to me through the mail for 8/28/2008! I did not show up! Then they sent me an appointment by mail for Sep 25, 2008. I did not show up, for this government run healthcare! That the U.S. government has assigned to Military Officers/Doctors, of a 3rd world country. The lowest bidder....

I Did Not Show Up, Would you?
We Have All Been Betrayed!

Chapter 16

SUSPICIOUS ACTIVITY

-----Original Message-----
From: Michael <michael.james44@sapo.pt>
To: [redacted]<[redacted]>
Sent: Wed, 2 Jul 2008 7:39 pm
Subject: [redacted]***<----my full name!!!!***

Dear [redacted], ***<----my full name!!!!***

I am Mr. Michael James, the Credit Manager of Clydesdale bank International in the United Kingdom. I have a proposal to discuss about a late, potential family member who had a Fixed account in Clydesdale bank online, Please contact me via email or call +44-702-401-6707 if you are interested to know more.

Regards,
Mr.Michael James.
Tel:+44-702-401-6707
Fax:+44-709-287-9730
Clydesdale bank,United Kingdom Laison Office.

<----my full name!!!! **Reader:** If it is a scam email, that is sent to everyone on a long list of email addresses. There is no way, in my mind, that they could know my full name.

This suggests that there is someone at one of the locations I have been sending SALUTE reports to. Who is leaking info and my contact info as well!

When I had seen on FOX News, that Senator McCain was in the country of Colombia on July 2, 2008, I was sure, in part, that it had to do with my "SALUTE REPORT" concerning my encounter with "Osama bin Laden." And, I wondered if that was where he was being held and water boarded!

However, then the next day, when the news broke about the hostage rescue, I was sure that Senator McCain had given the Colombian government the info needed to find the hostages.

I have never stopped sending in SALUTE REPORTS, and sit around watching the world change for the better after each report. I am sure that my life has not been taken, because of the quick action of Senator McCain, and his office, after each SALUTE REPORT.

This next email from Senator McCain's office is after my reporting about the Subj: **Prepare For An International Incidence!**

Subj: **Correspondence from Senator McCain**
Date: 7/3/2008 5:29:35 A.M. Taipei Standard Time
From: senator@mccain.senate.gov
To: [redacted]
Sent from the Internet (Details)

July 2, 2008

Mr. [redacted]
Dear Mr. [redacted]:

I wanted to take a moment to thank you for your recent correspondence. I appreciate your taking the time to assist me in keeping up to date. I found the material you referenced to be very informative.

I have passed along the information you forwarded to my foreign-relations and defense staff. I am confident that it will be useful. In the future, please do not hesitate to contact me on this or any other matter.

Again, thank you for the information. I certainly hope that you will continue to keep in contact on this and other matters.

Sincerely,
John McCain
United States Senator

JM/ab

That was the final correspondence that I have received from Senator McCain to date. In spite of the fact that, the situation concerning my disability compensation being turned off, has been resolved! The situation concerning my encounter with Osama bin Laden, continues on to date. Many times I have sent a message requesting info as to whether or not he was captured. I have never gotten a straight answer, **No admittance, no denial!**

During the months to follow, I sent a SALUTE REPORT to Senator McCain about every little thing that happened. Which is what I am putting into BOOK TWO…

During the first week of Feb 2009, I called home to Florida to find that the Social Security Administration had sent a letter to my Florida address giving me only 4 weeks to reply, or they would turn off my Social Security Compensation once again.

It had only been 6 months, and now I was going to have to go through the same fight all over! This brought out in me an overwhelming desire to climb up and out, of ever depending on the Social Security Administration and Government Run Health Care, ever again!

It is run like a dictatorship! I have had my fill, which pushed me into spilling the beans in this book!

What you will read next, is the message I sent to Senator McCain the day before I started writing this book. I hope to put it as the first page of Chapter 1, Book Two /

"Rats Patrol"…

Dear Senator McCain, Sir!

I am concerned about being followed so much.

As you know, I sent in a "SALUTE REPORT" to you on many circumstances over the past year and have been investigated to no end. During a few reports, my son and I reported being followed through the city of Olongapo Philippines. It has been obvious that Senator Gordon and or Mayor Gordon of this area have received Intel on this subject. As their people are not very good at working covertly… Their actions speak loudly…

At one time a report was sent through your contact form on your Gov website about suspicious activity on the street I live on, dealing with Islamic men wearing "orange shirts" and showing concern about my son, and or I, walking in front of their apartment.

This went on for a few months! We kept away as much as possible. We were even going the long way around, by use of another street. During one event, about 8:30 at night! As I walked with my son to catch a Jeepney for his ride home! An African man was standing on the sidewalk looking toward the vehicles that belonged to the Islamic men we had been trying to avoid.

It was dark (8:30pm), he was on the sidewalk in the shadow from the street light wearing very dark sunglasses. My son said "Dad, Dad, do you see that?" I said yes and to just keep walking. As we came closer to the African man (note: I'm saying African, not African American! I know the difference!) he turned his head toward us and seemed shocked to see us.

As we made our way past him and on to the end of the street, we stopped. I looked back and the man was gone.

A few days later, around 4pm in the afternoon, I was walking the same route alone. The African man was in the same place. He was again watching the vehicles that belonged to the Islamic men. Again he was shocked to see me. But, this time he turned into a doorway as I approached.

As I came to the doorway, I turned my head and looked to see if he was watching me. He was gone, nowhere in sight, and the door was the opening to a hall way. Which was about 50 feet long, much too far for him to have walked before I walked by? So I assumed he had run the length of the hallway as there were no doors on either side. Only a turn to the right at the far end which leads to apartments upstairs!

I never saw the man again, but had to walk by the Islamic men's apartment a time or two, so that I could get where I was going. Once, before I came within 100 meters, an African man in a white T-shirt walked toward me as I approached and said **"You will die!"**

Each time, I was given a dirty look by one of the Islamic men that were standing outside the apartment.

Once a man dressed as if he were in Afghanistan sat outside, obviously as a lookout! It seemed like a meeting of many was going on inside, as 4 visiting new SUV's were parked outside on the street…

The Islamic men have sense moved away, so we no longer worry when walking by that apartment.

But, today Feb 24, 2009 4:00pm to 5:00pm Philippine time! My wife and I took a Jeepney to the KFC not far from the "Victory Bus Terminal" but on the main street through town for a meal.

After we had eaten, we walked across the street to an ATM. Then back across the street again and on northward to the "Mart 1" department store to buy new socks for me.

As we entered, an African man around 45 years, 6ft, 190lbs, dark pants, white T-shirt with dark blue ring around the caller and hem of the sleeves end, dark blue base ball type cap, was 5 feet inside and turning toward us. He was shocked to look me in the eyes and became very angry.

My wife always looks down when something like this happens, but, I look forward, and act like, I don't know what's going on.

As I follow my wife on into the store, the African man puts out his right elbow and nearly hit my wife in the ear as she walked by. As I followed past him, his elbow came further out to block my way. So I turned sideways as I passed him and looked at the side of his face, noticing that he was watching the 2 guards at the entrants' door. His elbow pushed into the upper part of my stomach as I passed by. 10 steps further, we were walking up stairs to the second floor. So I turned to get a good look at what he was doing.

He was still watching the 2 guards, and now they were watching him. It was a stare down between them.

I realized he wanted a confrontation with me, but the 2 security guards were watching him.

Note: I do not believe he started out looking for a confrontation with me. He was only using a pretext. Meaning he was on foot watching us, and walked in the department store so that he could watch us walk by unnoticed. Then continue following us as we moved down the sidewalk.

Knowing this, I know that he most likely was not working alone.

Which brings me to this?

I request a team with fire arms.

Mission: Find who they work for. And if needed, take them out.

Senator McCain Sir, my being in danger is one thing. My wife and son being in danger is not expectable. Anyone thinking or stating that I am dreaming this all up is totally and completely wrong..

Sir, as my record proves, I am not a liar. And, I would never lie to you. After my encounter with "Osama Ben Laden" last year and my SALUTE REPORT to you about it and the reports that have followed. I have never been interviewed.

Even the most basic investigator will do an investigative interview with witnesses. I.e., myself, my wife, the man from immigration and the people that were working at the café where the encounter took place! To include looking over any cctv cameras in the area.

This has not taken place, and until it has,
the saga continues……

Person or person's working for Senator McCain, get this to him.

Senator McCain Sir, I realize you are very busy. And, I do not wish to be a thorn in your side. Can you please authorize an investigation as to how Senator Gordon and or his father Mayor Gordon here in the Philippines? Receives, and act's on messages, I have sent to you. And, why no interview with witnesses to my encounter with "Osama Ben Laden" has taken place!

Thank you for your time kind Sir!
Purple Heart [redacted]

Chapter 17

The Cover Up

Wow, while doing a Google Search on the Internet for this book, look what I found!

Obama administration tries to kill e-mail case
Updated February 22, 2009 05:05 AM http: // www philstar com / Article aspx?articleId = 42550& publication Sub CategoryId = 200

WASHINGTON **(AP)** - T*he Obama administration, siding with former President George W. Bush, is trying to kill a lawsuit that seeks to recover what could be millions of missing White House e-mails.*

Two advocacy groups suing the Executive Office of the President say that large amounts of White House e-mail documenting Bush's eight years in office may still be missing, and that the government must undertake an extensive recovery effort. They expressed disappointment that Obama's Justice Department is continuing the Bush administration's bid to get the lawsuits dismissed.

During its first term, the Bush White House failed to install electronic record-keeping for e-mail when it switched to a new system, resulting in millions of messages that could not be found.

The Bush White House discovered the problem in 2005 and rejected a proposed solution.

Recently, the Bush White House said it had located 14 million e-mails that were misplaced and that the White House had restored hundreds of thousands of other e-mails from computer backup tapes.

The steps the White House took are inadequate, one of the two groups, the National Security Archive, told a federal judge in court papers filed Friday.

"We do not know how many more e-mails could be restored but have not been, because defendants have not looked," the National Security Archive said in the court papers.

"The new administration seems no more eager than the last" to deal with the issue, said Anne Weismann, chief counsel for Citizens for Responsibility and Ethics in Washington, the other group that sued the EOP.

The Executive Office of the President includes the president's immediate staff and many White House offices and agencies.

Tom Blanton, director of the National Security Archive, noted that President Barack Obama on his first full day in office called for greater transparency in government.

The Justice Department "apparently never got the message" from Obama, Blanton said.

The department defends the government when it is sued.

Please note: Take a look at the "CIA World Fact Book," on the internet concerning the Philippines. I was shocked that nothing was mentioned about the many political murders that took place in the Philippines during 2007. Nevertheless, I was encouraged when I had seen near the bottom of the background section, the sentences that said "The Philippine Government faces threats from three terrorist groups on the US Government's Foreign Terrorist Organization list, but in 2006 and 2007 scored some major successes in *capturing or killing key wanted terrorists*. Decades of Muslim insurgency in the southern Philippines have led to a peace accord with one group and on-again/off-again peace talks with another."

Reader; every time I think back over my encounter with "Osama bin Laden," questions arise in my mind!

1.) Was this all just coincidence? And, because many people I have talked to, question what I tell them. Was it a look alike? Was it one of his brothers, and the old woman was "Osama bin Laden" in drag?

NO!

I did prey before I came to the Philippines while still in the country of Panama! That, if at all possible, I would find "Osama bin Laden," if he was there!

2.) Was "Osama bin Laden" there to meet with someone from the office of immigration?

a. The man from immigration was late because he had to help a client! Was it "Osama bin Laden" and the people with him?

b. Had he just come from helping them, and this is why he wanted to go to those outside tables. As if, it was easily out of view from the office of immigration, and he could do business without anyone discovering that he was over charging me by 2,000 pesos.

3.) Did the man from immigration know that they were there, before we walked over to them?

a. If so, was he and or "Osama bin Laden" testing, to see if "Osama bin Laden" would be recognized?

b. Or, if so, was I being tested! I say this, because I believe I was tested in the country of Panama by KGB as someone that might be anti US Government. / You will read about that event later, maybe in book 2.

c. Or, did the man from immigration suspect them and wanted to see if I would confirm his suspicions?

4.) Did "Osama bin Laden" get killed after my "SALUTE REPORT?" And, now a big cover up, is underway! Is the new "Bin Laden" tape's all fakes? Was he water boarded, and purposely breathed in water to commit suicide? Is that why, "Saudi Arabia" was so angry with "President Bush" when he was in "Saudi Arabia?" And, is that why, Obama is trying to have Bush prosecuted? If "President Bush" did not tell them (Saudi Arabia) about my "SALUTE REPORT," who did? Was it "Obama" when he was in the Caribbean where he knew he could call overseas and not be listened in on by the "Office of Homeland Security?"

After receiving my "SALUTE REPORT,"

They did not interview me!

They did not interview my wife!

They did not interview the man from immigration!

They did not interview the office of immigration!

They did not interview the people working in the coffee shop!

They did not check any CCTV cameras in the area!

How could they have not gotten him, and not have any questions about my reported encounter with him / "Osama bin Laden?" If they didn't get him, surely they would have questioned me! Alternatively, is this simply "Automatic Denial?" Because, if they want to cover up that they got him, they have to leave me in the dark and you as well!

These questions have never been answered! You will soon become very angered and understand why I have put them here!

"President Bush" was dancing on the front steps of the White House when Senator McCain, was visiting just after my "SALUTE REPORT" and his nomination as the Republican Presidential candidate!

Dancing?

Now that you know about my encounter with "Osama bin Laden" and my "SALUTE REPORT" to President Bush, which you read! You can see his dancing on the front steps of the White House, with the same perspective that I see it with!

What a great pretext! I just knew they had captured or killed "Osama bin Laden" and, I had earned the reward money!

After my "SALUTE REPORT," Senator McCain, went to Iraq, a few days later, the Iraqi Army led during their first major assault! During the same trip, Senator Clinton was having flash backs about phantom snipers. "Barack Hussein Obama" took a sudden unplanned (Cloak and Dagger) trip, to the Caribbean.

Did your eye's just open wide?

They should have!

Two plus Two equals Four!

Connect the dots: We had been betrayed; someone had told him <Obama> and he was worried! This is why he <Obama> went on a sudden unplanned (Cloak and Dagger) trip, to the Caribbean during that time.

All Americans know that you have to leave the USA, when you want to or need to, make an overseas phone call. So that the Office of Homeland Security or the FBI cannot legally listen in on the call!

The Governor of Illinois, Governor Blagojevich should have done like Obama and left the USA before he said anything over the phone. He would have never been discovered, if he had only thought it out!

That makes sense, as to why "Senator McCain" took such a sudden trip to Iraq. He would not take any chance of anyone listening in on the call and leaking the information. He would give it to the commanders in the field in person, behind closed doors.

Who told "Senator Obama" about my "SALUTE REPORT?"

Was it "Colin Powell?"

Was it someone in congress?

Was it the speaker?

Who did "Senator Obama" need to call overseas so badly that he needed to leave the USA to do it? And, why would he need to do it covertly?

Did your eye's just open wide?

They should have!

Two plus Two, equals Four!

Did he need to work against the capture of "Osama bin Laden," to insure that he could not lose the election?

Did he call "Saudi Arabia?"

If I did not mention this, I would be helping him get away with a crime against our people…

It's up to him, to disprove my theory, about why he took a sudden unplanned (Cloak and Dagger) trip, to the Caribbean. So that he could call the King of Saudi Arabia and get his help, from the capture of Osama bin Laden in the Philippines causing him to lose the election.

With all the problems bad people have given "Opera" over the years, surely she would have vomited if she had known what he had done while riding her good name!

This was all unfolding before my eyes!

Was he mindful, of the fact that he was working against me as well? On the other hand, did he even care?

Was it someone in the CIA or FBI that let him know the reason behind "Senator McCain's trip to Iraq?" I do not think so, they have always put America first!

"Senator Obama" was very happy after returning from that sudden unplanned trip to the Caribbean and nothing could break his strong belief that he was going to win the election from then on; he knew something we didn't know; something "Cloak and Dagger"!

Even something that "Joe Biden" did not know!

Alternatively, even "Hilary Clinton," if she had known, why "Senator Obama" had gone on this sudden trip out of the US, she would have come out punching like "Mohammad Ali!"

Did your eye's just open wide?

They should have!

Two plus Two, equals Four!

As I watched this unfold, I became increasingly offended!

All American's need to know! Because, if he worked against "Senator McCain and President Bush's efforts" after my "SALUTE REPORT;" and events do point this out! Then he is a trader on the level of "Benedict Arnold!"

"Who told him?"

Soon the people who know will start coming forward, and people in high places will begin their drag through hell! My turn at being dragged through hell has been over for a long time now!

Who cares if he and his office try to discredit me by saying that I am insane? All you have to do is take a look at the "Time Line" between "My SALUTE REPORT" and "What Took Place during the Days After," even until this very day, to see the truth!

For you, the knowledge of my "SALUTE REPORT," puts everything into a new and more enlightened perspective!

Your motto in life should now be, "TRUTH, GET SOME!"

5.) When President Bush visited Australia after my SALUTE REPORT, he had a closed door meeting with the Prime Minister of Australia. After the meeting, the Prime Minister of Australia was videoed giving President Bush a full "HAND SALUTE" from across the room.

Knowing what I know and what you now know, which is, "President Bush" read my "SALUTE REPORT" concerning my encounter with "Osama bin Laden." We can believe together, that the "Prime Minister of Australia" had been told about the "SALUTE REPORT" by "President Bush" while in the closed door meeting. And, now the "Prime Minister of Australia" was giving "President Bush" a "HAND SALUTE" out of respect from being told about it (April 2008). Or, even that it was "A Go," that Australia would be happy to supply a covert location inside Australia to keep "Osama bin Laden" hidden at, while being water boarded.

In Australia, where could the CIA hold "Osama bin Laden" with the permission of the "Prime Minister of Australia?"

Go to Google Earth, Australia: zoom in at 23 degrees 35'33.24" S by 134 degrees 29'36.13" E, now tell me; does that look like a spot where helicopters have been landing often? Now move over to the right at 23 degrees 35' S by 134 degrees 30'55.37" E. That's a nice airplane landing strip, isn't it!

Australia is not my enemy, nor am I there's! So when they discredit this info, I understand completely, and rightfully so!

6.) Then when it came across the news channels that the US government did not want to give up "Water Boarding" as a way to get information out of a terrorist, because they needed it for "Osama bin Laden?" I believed it was a sure thing that my "SALUTE REPORT" helped them find him. They had him or were sure they would have him soon. Which meant that I had earned the reward, right?

7.) When "Senator McCain," went to Columbia and then left the day before they suddenly landed on top of and then freed the FRAC prisoners. It looked like he was there giving them the info needed for the successful rescue operation and to ask for a safe place to keep "Osama bin Laden" in return! On the other hand, that maybe he was there already or would be soon, and the info was a way of helping them, the same way they were going to be helping us.

8.) When the "President of the Philippines" visited the Whitehouse a few months after my "SALUTE REPORT," she and "President Bush" were shown at a photo op on the international news. "President Bush" said that the "President of the Philippines" was working with or helping the US in the "War on Terror". However, "President Bush" had a very unhappy look on his face. What bad news did she give him? Was it the news about the death of "Osama bin Laden" while he was being water boarded? Alternatively, was it news that he had gotten away?

Surely, she reported who it was, that had been caught at the International airport in Manila, on or around Mother's Day, during 2008, trying to get out of the country! Remember that the Philippines cut and run during the Iraq war! The content of their character has proven that they are not reliable allies. After my "SALUTE REPORT," the entire Philippine Military had to go for a complete loyalty test and many lost their jobs or were arrested.

9.) What you are reading, is not a conspiracy! What this is, is an eyewitness account! What you are reading, I have seen with my own eyes and experienced with my own being!

10.) The US Navy has been here in the Philippines 100% of the time sense my "SALUTE REPORT" concerning my encounter with "Osama bin Laden." And just a few days after my "SALUTE REPORT," was sent to "Senator McCain," a C-130 did a low flight "BUZZ" over the top of my home here in the Philippines! I took it as a CIA heads up, to let me know that they were all over my "SALUTE REPORT" concerning my encounter with "Osama bin Laden!" Gee, I love those people in the CIA; or was it the people of the US Navy's Intelligence corp.'s? They are some kind of good at what they do!

11.) President Bush, while visiting Saudi Arabia after my "SALUTE REPORT," ask the King for the increase of oil production by his country to help bring down the price of gasoline. The King refused! Had "President Bush," informed him about the capture of "Osama bin Laden" and it pissed him off, or embarrassed him in front of the "Bin Laden" family; or had someone else already told him long before?

Then the hunt for "Osama bin Laden' settled down! It was as if "Saudi Arabia" had turned against the US for capturing "Osama bin Laden" or killing him! And, just like they did during the Vietnam War, did an oil embargo to control, or affect the outcome of the Vietnam War, and now of the US presidential race. Always remember, that during the Gulf War of 1990 and 91, all Saudi Arabia would give members of the 82nd Airborne to eat, was one "Hot Dog Bun" three times a day.

Do not let President "Barack Hussein Obama" get away with investigating his own office again, like, he did when the news broke about the Governor of Illinois, Governor Blagojevich.

Do a Google search on the internet for everything that happened in the "Philippines" during 2007 and 2008 even until this very day? Pay special attention, to anything and everything after the time of my "SALUTE REPORT" to include President Bush's activities along with "Senator McCain's!" Do not leave out "Barrack Hussein Obama!" And do not fall, for the race card issue, like the world did during the first trial of "OJ Simpson!"

The color of your skin is no excuse for bad behavior! The Governor of Illinois, Governor Blagojevich should have done like Obama and left the USA before he said anything over the phone. He would have never been discovered, if he had only thought it out!

Look at the time line of events; when "Obama" said to "McCain" about "Osama bin Laden," "You don't have him, do you?" McCain had a look of betrayal!

And never forget to do a website archive search!

Go to http // web archive org/ and do a search of the news channels during the dates you need to find info on!

Meanwhile, on the O'Reilly Factor, Dennis Miller is making me smile!

You now have a spot light, while you walk through the darkness! And your perspective on the days to follow is much clearer!

You should be doing a "Google Search" right now and forever!

Call "Lion Investigation Academy" (**Toll Free: 1 (877) PI School)** and, get started with a "Private Investigation Course" right now, this very minute! It can and will, help you in your day to day life!

And, always remember how to, **SALUTE REPORT**!

Size: 1

Activity: Suspicious activity; while researching on the internet for my book "SITUATION RESOLVED", I did a Google search for the date and time that "Bill Richardson" went on vacation in the Caribbean. To see if, it is the same time that Obama went to the Caribbean, while McCain was in Iraq! To find out who might have told Obama about my SALUTE REPORT concerning Bin Laden!

Google brings up http // www google com ph / search?hl=en&q = Mr. + Richardson + had + been + scheduled + to + go + on + vacation + in + the + Caribbean + in + march + 2008&btnG=Search&meta=

1. Bill *Richardson* Makes A Hard Choice | PoliticalDiner.Com

Posted on 22 *March 2008* by chef. I woke up this morning to a loud sound. .... because *Mr. Richardson had been scheduled to go on vacation in the Caribbean. ...*

politicaldiner.com/?p=36 - 46k - Cached - Similar pages

Once I click on the link "politicaldiner.com/?p=36" I am taken to a page that says "Your Website Has Been Suspended!"

So I went to the website achieves at http // web archive org / web / * / http // politicaldiner com /.

There, all history of www.politicaldiner.com has been deleted all the way back into 2002…

Location: Internet website archive's at http // web archive org /

Uniform: none

Time: 4:10 PM, May 20, 2009 Philippine time.

Equipment: 1 home computer with internet connection!

Sing out, ole news man, sing out!
He was hiding in the open..
Sing out, Ole news man, sing out!
He was living in modern luxury,
Sing out!
Let his al-Qiada now, sing out..
He was living in luxury with women,
Dressed like an American on vacation,
While his fighters lived in caves, and died for his cause.
Sing out...

It is time that all Americans demanded an investigation!

Question's to bring up:

1.) Why have I never been interviewed in person, concerning my reported encounter with Osama Ben Laden?

a.) Even the most basic investigation has a question and answer interview with the witness and or witnesses.

2.) Why were the CIA and or US Government not helpful in the recovery of the photos?

3.) Why is the Philippine Government being rewarded, with a 138 million dollar grant? In addition, the American people are being told that it is for the Philippine WWII veterans!

4.) Why did Obama go on vacation during the period of the investigation? Note: Clinton's flashbacks about a phantom sniper were during the same time..

I want a fully transparent investigation by the US Congress as to how many US Embassies around the world are manned by non US citizens. How many in total employed at US Embassies around the worlds, who are not US citizens? How much in total are they paid? In addition, why? (With as many American's out of work as there are. Why is the government not employing more of them in our embassies?)

April 9, 2009 on FOX news interview, Father Morris: *Suffering can be a great time, but only for humble people! Anything that causes accountability, is not only moral, it is beneficial!*

Trust me on this one, all that's needed to change the world, is for all believers in Christ Jesus. To come together at the same time, around the world, and bring forth from deep within their soul, out over their tongue, through their lips, a truly heartfelt thunderous "JESUS!"

Not for the person next to you or across the street to hear, but to release your total faith and surrender to "JESUS." He will take charge!

Then, anytime in your life that you find yourself faced with a situation, where you must trust in your fellow man, repeat the Lord's Prayer, then let go of it.

If you feel that your specific need is not known by God, and you must speak it. Then go ahead and ask, but be prepared! The Lord is not a slave to do your bidding. He has free will, above all free will. He has the right to say no and refuse you.

So at the end of the Lord's Prayer, be sure to say thank you, thank you. And, thank you again. Then accept what follows.

He might say, "No!" Or, he might just say "Yes, I'd like to do that, but first I want you to stop killing your baby's by aborting them." Or he might say, "You know what, I really don't agree with gay marriage! So let's take care of that first!" Or, "Yes, I'll be happy to do that for you. Can you clean out the porn from your home, so that I might visit without needing to cast out demons? It's your choice!"

What I'm trying to point out here, is, we all need to clean out our own home. And, we need to do it today and every day. We all need to be rats, ratting on anything and everybody that's doing wrong in our world. We dirtied it; we need to clean it up.

So, the next time you witness someone selling an illegal drug or breaking the law. Even if your neighbor's child is sneaking out a window late at night, rat on them, by going to a law enforcement website and submitting a **"SALUTE REPORT."**

Be the best volunteer undercover operative that you can be! Do this knowing that nobody knows about it, unless you tell them?

In closing, I was thinking to myself just the other day: Who am I, that a great man like "Senator John McCain," would take notice off, and show respect toward?

Before I had written this book, I had gone through life trying to find the person I use to be, I just could not do it, until now!

America, if you would choose to pray for me;
I have just one request!

Put Silver Wings,
On My Sons Chest!
Make Him One,
Of America's Best!

THE END

Be prepared for Book 2, "RAT's PATROL,"

The Saga Continued....

AUTHORS' NOTE:

History has proven;

The world is filled with terrorists, drug cartels, drug dealers, drug users, predating pedophiles, stalkers, rapists, good cops, corrupted cops...

The list may go on to even more monstrous cases: fathers and mothers killing their babies, entire families slaughtered in their own home, or both parents murdered and their young children disappeared with no trace left for the police to track...

Law Enforcement Officers not using K-9 search dogs for 2 days, while young children die in the trunk of a car right under the officers' noses, in the yard of their own house...

Sounds like a movie script. It's not!

But if you watch any of the major news channels, you know it's not! And another sad "if" - if you're in the United States of America, you know - the laws protect anyone accused of a crime. Then you realize, it is your word against theirs…

That being said, still I must speak up concerning what I have seen with my own eye's, experienced with my own being! For if I witnessed a hit and run accident on a highway, then never reported it. I would be aiding the perpetrator, allowing them to remain at large. Going about their life, as if it never happened…

Let this eyewitness account, through-out this book, be a testimony under oath, in a court of law!

Even so; As you can see throughout this book, I have been betrayed by those I trusted many times during my life... Now once again I have been betrayed! Not by someone close to me or someone that I work for or even someone I work with. Nevertheless, by members of my own government…

How many of you can say that? Really, that many...

Well then, you should have no problem understanding when I state; "We have all been betrayed!"

Further, notice that I use the term 2+2 equals four! But then, I am from the old school. The term that is used more commonly in this day and time is the term "Connect the dots!"

Either way you want to rationalize this; what you need to know is, "The Obama Administration will not allow the emails from the Bush Administration to become public until after he (Obama) has a chance to run for his second term in office as President of the United States of America.

Connect the dots on that one!

I take this personally, and feel betrayed! As it is very important that you know beyond the shadow of a doubt. That the copies of the emails to President Bush, concerning my encounter with Osama bin Laden inside this book. Which you have read; are absolutely authentic…

So, what I watched on TV told me that Senator Obama was shook up so badly that he suddenly left the country during the campaign. When covert operations here in the Philippines, point to the US Government and the Philippine Government. Acting on the information, I provided them in my SALUTE Reports…

Now add 2 and 2! Connect the dots!

Why was the Philippines suddenly given such a large amount of money by the US Government? Right after the covert operations which took place during the month after my SALUTE Report about Osama bin Laden…

Now connect the dots on this; right after the covert operations in the Philippines slowed and the entire Philippine military went though it's restructuration... President Bush made a stop in Australia. If you watched it, you had seen the Prime Minister of Australia. Who was elected after a campaign which he won by bashing President Bush... Then after a closed door meeting with President Bush, the next time he sees him, they are across the room from each other and the Prime Minister of Australia gives President Bush a full hand SALUTE...

What do you get when you connect the dots on that one?

I get a very possible; the Prime Minister of Australia asked President Bush! How did you know where to find him (Osama bin Laden)? President Bush answers; Through the information I received in a SALUTE report!

Now connect these dots; for many months during 2008, after I sent in the first SALUTE report. Zero messages were received from Osama bin Laden. Then when they started up again, they used old still photos of him and an altered voice...

Hey, and do not forget the SIM card with the photos!

2+2=4; who needs to recover photos of Osama bin Laden, when you already have captured him...

Inclosing; if the people of the USA, do not stand with me, and demand that an investigation into my email SALUTE reports. And what action was taken with the information they provided. We Americans, will go through another presidential election. Without knowing the total truth...

Because, alone; all I can get out of the US Government, is "No Admittance, No Denial!"

Mr Cloak

Lightning Source UK Ltd.
Milton Keynes UK
UKOW051534250412

191444UK00001B/120/P

9 780982 737101